TECHNICAL REPORT

Employing Land-Based Anti-Ship Missiles in the Western Pacific

Terrence K. Kelly • Anthony Atler • Todd Nichols • Lloyd Thrall

Prepared for the United States Army

The research described in this report was sponsored by the United States Army under Contract No. W74V8H-06-C-0001. The findings and views expressed in this report are those of the authors and do not necessarily reflect the views of the Army or the U.S. Department of Defense.

Library of Congress Cataloging-in-Publication Data is available for this publication.

ISBN 978-0-8330-7792-9

The RAND Corporation is a nonprofit institution that helps improve policy and decisionmaking through research and analysis. RAND's publications do not necessarily reflect the opinions of its research clients and sponsors.

Support RAND—make a tax-deductible charitable contribution at www.rand.org/giving/contribute.html

RAND OFFICES

SANTA MONICA, CA • WASHINGTON, DC

PITTSBURGH, PA • NEW ORLEANS, LA • JACKSON, MS • BOSTON, MA

DOHA, QA • CAMBRIDGE, UK • BRUSSELS, BE

www.rand.org

Preface

This report was produced as part of the RAND Arroyo Center project "Preserving Army Capabilities in a Time of Downsizing," which was designed to help the U.S. Army's Office of the Deputy Chief of Staff for Programs (G-8) as it considers the future capabilities needed in the Army.

The purpose of this report is twofold. First and most directly, it puts forward a concept for how land forces could play a significant role in a conflict with China in the U.S. Pacific Command area of operation should Chinese aggression threaten U.S. allies or interests. Second, and perhaps more importantly, it illustrates that analysts need to think about what is needed to deter such a conflict and how it could be prosecuted from "first principles." We hope that such an analysis will help the U.S. Department of Defense make decisions about force structure and operational concepts that advance U.S. interests, rather than simply looking at how to modify current capabilities. The findings and views expressed here are those of the authors and do not necessarily reflect the views of the U.S. Army or the U.S. Department of Defense.

This research was sponsored by Volney J. Warner, director of the Army Quadrennial Defense Review Office, U.S. Army G-8, and conducted within RAND Arroyo Center. RAND Arroyo Center, part of the RAND Corporation, is a federally funded research and development center sponsored by the United States Army.

The Project Unique Identification Code (PUIC) for the project that produced this document is RAN126175.

Questions and comments about this research are welcome and may be directed to the lead author, Terrence K. Kelly, at 412-683-2300, x4905, or Terrence_Kelly@rand.org.

For more information on RAND Arroyo Center, contact the Director of Operations (telephone 310-393-0411, x6419; fax 310-451-6952; email Marcy_Agmon@rand.org), or visit Arroyo's website at http://www.rand.org/ard.

Contents

Figures

Tables

Summary

In his strategic defense guidance of January 2012, President Obama declared that U.S. economic and security interests are "inextricably linked to the developments in the arc extending from the Western Pacific and East Asia into the Indian Ocean and South Asia."[1] In doing so, he shifted the U.S. focus to the Indian and Pacific Oceans and provided a set of precepts that will shape the future orientation of the joint force.

This shift in strategic priorities to East Asia was preceded by a growing literature about threats to the ability of the United States to project and sustain power there. Over the past several years, some strategists have argued that China is shifting the balance of power in the Western Pacific in its favor, in large part by fielding anti-access weapons that could threaten U.S. and allied access to vital areas of interest.[2] Others have argued that such innovations have lowered the costs of anti-access capabilities such that regional actors can contest "America's 60-year-old dominance over the global commons and its ability to maintain their openness."[3]

As a result, new concepts such as "AirSea Battle" are being developed to "set the conditions at the operational level to sustain a stable, favorable conventional military balance throughout the Western Pacific region."[4] In general terms, AirSea Battle envisions integrated Air Force and Navy operational concepts to mitigate missile threats to U.S. bases; correct imbalances in strike capabilities; enhance undersea operations; offset the vulnerabilities of space-based command and control (C2) and intelligence, surveillance, and reconnaissance systems; increase interoperability; and enhance electronic and cyber warfare capabilities.[5] It would do so by improving the "integration of air, land, naval, space, and cyberspace forces to . . . deter and, if necessary, defeat an adversary employing sophisticated anti-access/area-denial capabilities."[6]

Although land-based systems feature prominently in China's anti-access/area-denial (A2AD) capabilities, comparatively little work has been done to define the land-based support capabilities featured in the AirSea Battle debate. Such capabilities may prove potent and

---

1 U.S. Department of Defense, *Sustaining U.S. Global Leadership: Priorities for 21st Century Defense*, Washington, D.C., January 2012a, p. 2.

2 See, for example, Andrew F. Krepinevich, *Why AirSea Battle?* Washington, D.C.: Center for Strategic and Budgetary Assessments, 2010, pp. vii–viii, 13.

3 Abraham M. Denmark and James Mulvenon, eds., *Contested Commons: The Future of American Power in a Multipolar World*, Washington, D.C.: Center for a New American Security, January 2010, p. 6.

4 Jan van Tol, Mark Gunzinger, Andrew F. Krepinevich, and Jim Thomas, *AirSea Battle: A Point-of-Departure Operational Concept*, Washington, D.C.: Center for Strategic and Budgetary Assessments, May 18, 2010, p. xi.

5 van Tol et al., 2010, p. xiv.

6 U.S. Department of Defense, *Joint Operational Access Concept*, version 1.0, Washington, D.C., January 17, 2012b.

inexpensive joint force multipliers. One such complementary approach is to develop concepts that employ the same inexpensive anti-access technologies to significantly raise the cost of a conflict for China and, should deterrence fail, to limit China's ability to inflict damage off the Asian mainland.

Approach

This report explores one such option: using ground-based anti-ship missiles (ASMs) as part of a U.S. A2AD strategy.[7] We note that if the U.S. military had such capabilities, it could use them in a host of ways, ranging from security cooperation initiatives to help regional friends and allies improve their own anti-access capabilities to using them to interdict warships or (if supplemented by other assets) help form a full blockade in times of war. We make no claim of having analyzed the strategic implications of deploying or employing anti-ship missile capabilities but, rather, seek only to demonstrate what is possible using existing capabilities and comment on possible contributions they could make should the United States adopt an A2AD strategy of its own.

To determine what is possible and to illustrate potential capabilities, we conducted a missile-by-missile comparison of 45 current anti-ship cruise missiles. These missiles are popular with armies in the region; China, Indonesia, Malaysia, Vietnam, and Brunei are all believed to possess multiple types of ground-launched ASMs (and some possess missiles that are fired from other platforms as well). Appendix A of this report describes some of the missiles that we considered.

We assessed the likely effectiveness of land-based ASMs by exploring the technical potential and possible impact of a U.S. anti-access strategy that could challenge Chinese maritime freedom of action should China choose to use force against its island neighbors. In our research approach, we assumed that the ability to cut off Chinese seaborne access beyond the first island chain would serve as a major deterrent, and would have a significant effect on China's ability to attack its overseas neighbors and wage a prolonged war. Furthermore, this capability does not require the permanent stationing of assets in the Western Pacific, and, as such, is not presented as part of an effort to contain China. Rather, it should be seen as a capability that could be used if China initiated a conflict. Finally, while U.S. Pacific Command will lead efforts to develop and execute military strategy in the Pacific, this report addresses the potential for U.S. land forces to significantly contribute to its efforts.

To illustrate the potential of land-based ASMs while acknowledging that a land-based-only approach would not be practical, this report examines the possibility of cutting off Chinese sea routes using land-based ASMs only. Not only would this have a significant effect on China's ability to project power, but it would also vastly expand the set of military problems that the People's Liberation Army (PLA) would face should it consider initiating a conflict with its neighbors or U.S. partner nations. Specifically, because these missile systems are relatively easy to operate and are strategically and tactically mobile (i.e., they are not fixed targets), the PLA would have to search across a huge number of locations and have assets within range to

[7] The utility of quickly deployable ASMs is not limited to Western Pacific contingencies or to China; they could be used in a range of scenarios in which the United States wished to limit, deter, or complicate an adversary's power projection potential.

locate and interdict them. Missile systems that could be placed in many locations over thousands of miles of island chains would significantly dilute the effectiveness of PLA missile and air forces.

It is important to point out that a comprehensive analysis of the regional political, economic, and military dimensions of these operations is critical but beyond the scope of this report. As a result, this report does not try to analyze how fielding ASMs would affect these strategic factors. Rather, it focuses on exploring the tactical and operational feasibility of ASM employment, including some observations on procurement and logistics. To this end, our approach assumed a wartime strategic context in which limiting Chinese maritime freedom of action would be important and in which a U.S.-led coalition decided to institute "far blockade." We also assumed that some regional states would be supportive of such a course of action in the context of a wider war, as their cooperation is important to our demonstration.

Finally, this report's purpose is to illustrate capabilities, not to make detailed recommendations about force structure or doctrine. Importantly, our intent is as much to encourage strategists to think of new approaches as it is to propose that the U.S. military consider developing a capability. This is but one such approach; there are others that should also be considered.

A Joint Approach

A land-based ASM capability would be relatively easy to create in the U.S. armed forces and could be seen as a 21st-century extension of the Army's earlier coastal defense role. It would need to operate as part of a joint effort. In general, it would require access to other services' (and perhaps national) sensor systems capable of identifying targets to engage, a C2 system that can receive and act on this information, and firing batteries that can respond to this C2 architecture. In this case, the range of these assets must span all passages through the straits that provide access to the seas surrounding China. If such a land-based ASM capability were to be used as part of a blockade, it would also have to be paired with assets that could challenge and board commercial ships, such as rotary-wing aircraft or partner-nation navies and coast guards. If allied firing platforms were preferred to U.S. ones, they would similarly have to be integrated into such a C2 architecture. In this report, we provide a general description of the elements of such a system while noting that more analysis would be required to produce a complete operational concept.

Feasibility of a Land-Based ASM Blockade

To illustrate the possibility of ASM employment, we built a "far blockade" of chokepoints in the Asia-Pacific region in phases.

Straits of Malacca, Sunda, and Lombok

The Strait of Malacca is both narrow and of significant strategic importance. Both Indonesia and Malaysia have robust arsenals of medium-range ASMs that, if these countries were willing, could effectively engage targets anywhere along the strait's approximately 730-km length. A coalition with access to these countries could similarly contribute. A concerted effort to close the Strait of Malacca using ASMs of medium capability would be difficult to defeat without employing land forces to locate these missile systems, which are mobile and relatively small in size.[8]

Longer-range ASMs would put ships under threat from missile batteries for even more time. For example, Indonesia's C-802 ASM, a version of the Chinese YJ-2 with a range of at least 120 km, is the farthest-reaching ground-launched ASM in the region. Other longer-range ASMs on the world market could effectively cover more than 1,200 km of the Strait of Malacca and the sea approaches around it. Finally, the BrahMos PJ-10, developed and produced jointly by India and Russia, could extend this coverage to approximately 1,500 km.[9]

A coalition's ability to deny China the use of the Strait of Malacca would not amount to a blockade, however. Ships coming from the Indian Ocean could simply use the next-closest waterways, the Sunda and Lombok straits. However, the narrowness of these passages means that they could be easily covered with short-range missiles as well.

Japan, Taiwan, and the Philippines

Should Taiwan and Japan be involved in a future scenario, ASM-based threats emanating from their territory would offer another capability to limit maritime freedom of action. For example, ground-launched ASMs located in Taiwan with a range of no more than 100 km, along with missiles with an effective range of 200 km in Okinawa, could effectively cover all naval traffic south of Okinawa. Another possibility is to position missiles with a 200-km range solely on the Ryukyu Islands, which would also effectively close the area south of Okinawa. The area between Okinawa and mainland Japan could be effectively covered by ASMs with a 100-km range in Japanese territory alone.

An implied task of this operational concept is to prevent Chinese forces from capturing these strategic islands. While this would be a joint and coalition force operation, it is useful to note that the very same ground-based A2AD systems used for the blockade would play a key role in these operations by targeting amphibious forces.

Similarly, the Luzon Strait between the Philippines and Taiwan, as well as the waterways between the Philippines and Borneo, could be covered by 100-km-ranging missiles positioned in the Philippines, Taiwan, and Malaysia or by 200-km-ranging missiles (in the case of Taiwan) or even shorter-range missiles (in the case of Borneo) fired solely from Philippine islands. The closure of these areas would significantly limit all naval activity, but more strategic depth could be achieved by also denying transit through the waters between Australia and Indonesia. While such a move is not a necessity, the requirements would include the use of cruise missiles with a range of approximately 300 km (such as the BrahMos PJ-10) positioned in both Australian and Indonesian territory.

Japan and South Korea

China may also wish to transit PLA naval vessels between Japan and South Korea via the Korea Strait. In such a scenario, ASMs with a range of 200 km could be launched from either Japanese or South Korean territory (or 100-km-ranging missiles could be launched from both sides or from the Japanese island located near the middle of this strait). However, as in the Strait of Malacca, operational flexibility and system survivability would increase with the use of both

---

[8] For detailed geospatial depictions of how ASMs could shut down all shipping routes to China, see Appendix B.

[9] The BrahMos PJ-10 is also one of only a few supersonic cruise missiles. Allegedly, a second version is being developed by India and Russia that is likely to be even faster than the original.

sides of this chokepoint. The full effect of the ASM deployments described here is illustrated in Figure S.1.

Logistical and Procurement Considerations

The deployment of ground-based ASMs could be quite flexible, permitting them to serve as a deterrent without requiring them to be permanently stationed in areas that the Chinese would see as threatening. Stationing ASM forces in the region would needlessly threaten and provoke China, as well as damage U.S. efforts to cooperate with China. Furthermore, host-nation access would be critical for employing ASMs and might not be forthcoming short of a conflict with China. However, should China use or threaten to use force against U.S. allies or partners in the region, the United States might want such assets available. As a result, it would need to be able to rapidly move ASMs into the region from U.S. territory or from other prepositioned stocks in Asia.

Figure S.1
Potential Chokepoint Engagement Areas for Ground-Launched Anti-Ship Missiles in Partner Nations

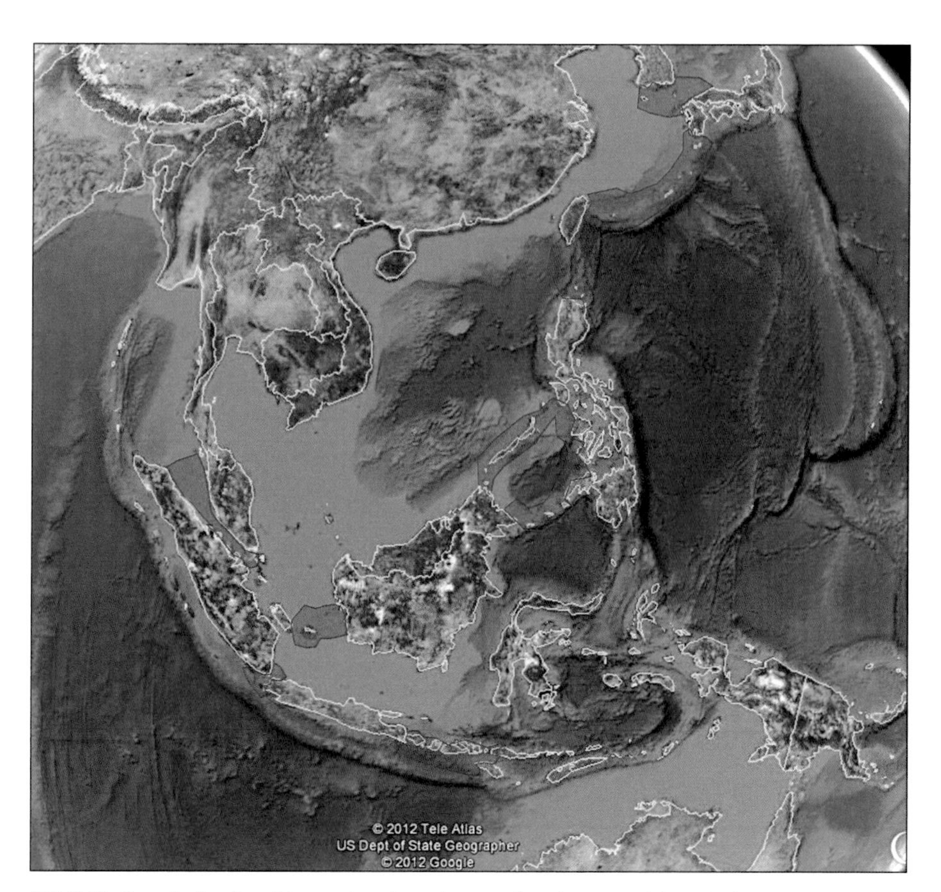

SOURCE: Google Earth, with overlays based on authors' geospatial analysis.
NOTE: Red areas are locations where access is denied by ASMs.
RAND *TR1321-S.1*

Many of these ASMs can be fired from a multitude of platforms and thus can be integrated with existing material and tactical requirements. The ability to transfer these missile systems to multiple platforms and deploy them from a number of vehicles with differing dimensions adds to their flexibility. However, this also makes it more difficult to determine specific mobilization and employment requirements.

One method to generalize these requirements is to review analogous systems and current U.S. mobilization methods. For this analysis, we used as a comparison the U.S. Patriot missile, which is longer, wider, and heavier than nearly all the cruise missiles considered here. The Army has determined the Patriot's minimum engagement package and identified a standard loading plan for C-5 and C-17 aircraft, so it can serve as a beginning point for planning estimates. The actual loads for any given mission would naturally be slightly different, as they would depend on mission-specific characteristics. The U.S. Army has established that the Patriot's minimum engagement package—which consists of two launchers, each with four missile canisters, eight total resupply missile canisters, radar and C2 systems, and all the personnel and equipment needed to fully operate the system—could be delivered with the use of five C-5s or seven C-17s.[10] Fast boats could also deliver these assets. As such, getting ASM systems into place during a crisis should be straightforward.[11]

With respect to procurement, the global market for anti-ship cruise missiles is wide-reaching and complex. With dozens of missiles available from nearly as many manufacturers and countries, there are a number of avenues through which one may procure missiles. The missiles highlighted in this report were chosen because of their capabilities and assumed availability. While cost information is available for all of the systems we considered, for our purposes, it is sufficient to note that missiles of these types are widely available for purchase and that creating a force structure to employ them would not require a major research and development effort.

Defense Relations and the Potential for Building Partner Capacity

Most of the nations upon which the United States would rely for access in this concept are strong partners or allies. However, Indonesia is arguably the most important for this strategy and has not traditionally been a close U.S. partner. Furthermore, while Indonesia currently accepts security assistance from the United States, it also is developing stronger relations with China. Building partnerships and, more importantly, persuading countries such as Indonesia to allow the use of ASMs on their territory may be the biggest challenges to carrying out the strategy outlined in this analysis. In a strategic context short of a direct conflict involving these countries and China, such assent may be difficult to attain because it would pose significant risks for the countries that agreed to cooperate.

That said, developing U.S. experience with ASMs would create opportunities for security cooperation with several Asian nations. Security cooperation is a mainstay of U.S. efforts to increase the capacity of partner nations, ensure access to territory, and influence other nations' behavior. Given the importance of the first island chain, it is no surprise that most nations there have these systems. Whether they can employ them effectively and whether they would

---

[10] Headquarters, U. S. Department of the Army, *Patriot Battalion and Battery Operations*, Washington, D.C., Field Manual 3-01.85, July 2010.

[11] We note, however, that they would also require other support systems, such as for security and logistics, which would add to the strategic lift requirement.

do so as part of a coalition effort are important questions. Yet, because the U.S. military does not have such systems, it is currently limited in how it can help build partner capabilities for their use. As a result, it may not be able to adequately influence the plans of allies and partners to deploy and employ them in concert with U.S. plans and efforts if they were needed to respond to Chinese threats.

Toward an Air-Sea-Land Concept

The Navy and Air Force may currently possess the capacity to contest Chinese maritime freedom of action in Asia without land forces. However, doing so would require using expensive systems that would, if successfully targeted by Chinese forces, be difficult to replace. An inexpensive truck-mounted missile launcher in an Indonesian jungle is considerably more difficult to locate and attack than an expensive naval warship patrolling the approaches to the Strait of Malacca—and yet both could contribute to blockade objectives. Furthermore, the demand for naval assets to control the sea lines of communication to U.S. bases in the Western Pacific and perform other missions in times of conflict would be significant. Land-based ASMs could help relieve some of these demands on the Navy (and Air Force). Additionally, positioning many ASM systems throughout the first island chain would very significantly increase the PLA's targeting requirements, stressing its C2 systems and causing it to spread valuable intelligence, targeting, and attack assets over many possible firing positions across an arch of islands that is thousands of miles long rather than focusing on a few well-defined targets. Arguably, this would significantly decrease the effectiveness of PLA anti-access assets and increase the effectiveness of other U.S. and coalition efforts.

The current AirSea Battle concept understandably places significant emphasis on the Navy and Air Force's capability to counter foreign A2AD threats. This report illustrates that creating an ASM capability in the U.S. ground forces could significantly dilute the A2AD threat and present a corresponding U.S. capability to an aggressor state that sought to project power over water. In short, developing and employing ASMs in the U.S. force structure would provide capabilities that could have a strategic effect.

Additionally, with such capabilities, the U.S. armed forces would be better prepared to work with Asian partners on developing their own ASM capabilities (as part of security cooperation and security force assistance efforts). Without such capabilities, it would be difficult to impart these skills and develop the relationships and access that come with these partnership activities.

Finally, capabilities such as those presented here will become increasingly accessible to nations and, perhaps, nonstate actors. Armed, unmanned systems (aerial and under water) could have similar effects to ASMs. Keeping these capabilities out of the hands of rogue actors will likely be an important task—one that could be used to build ties with China in the form of nonproliferation regimes, because both China and the United States would have a large stake in such efforts.

Conclusions

Land-based ASMs are readily available on the world's arms markets, inexpensive, and able to provide significant additional capabilities to U.S. forces. Their employment would require multinational and joint concepts and approaches, as well as support from other service assets, such as sensors, intelligence, and C2 systems. But the capabilities they could provide a multinational force would free up the Navy and Air Force for missions other than controlling maritime traffic (military or commercial) near land chokepoints. These capabilities would also significantly complicate the PLA's C2, intelligence, and targeting requirements and would raise the risks and cost of a conflict for China (and other nations that depend on maritime freedom of action or wish to project power overseas). Having such capabilities in the inventory would further U.S. efforts to provide security cooperation assistance to partner nations, could help deter conflict, and could contribute to victory in a future conflict by increasing flexibility and expanding the set of tools available to U.S. commanders to implement plans.

Acknowledgments

We would like to acknowledge the valuable insights and critiques provided by many RAND colleagues as we developed this analysis. We particularly thank Alan Vick and James Dobbins, who, along with several others, were kind enough to read and comment on our work. This report is better for their input.

Abbreviations

A2AD	anti-access/area denial
ASM	anti-ship missile
C2	command and control
ISR	intelligence, surveillance, and reconnaissance
PLA	People's Liberation Army
USPACOM	U.S. Pacific Command

Land-Based Anti-Ship Missiles in the Western Pacific

In his strategic defense guidance of January 2012, President Obama declared that U.S. economic and security interests are "inextricably linked to the developments in the arc extending from the Western Pacific and East Asia into the Indian Ocean and South Asia."[1] In doing so, he shifted the U.S. focus to the Indian and Pacific Oceans and provided a set of precepts that will shape the future orientation of the joint force.[2]

This shift in strategic priorities to East Asia was preceded by a growing literature about the threat to the ability of the United States to project and sustain power there. Over the past several years, some strategists have argued that China is shifting the balance of power in the Western Pacific in its favor, in large part by fielding anti-access weapons that could threaten U.S. and allied access to vital areas of interest.[3] Others have argued that globalization and technological innovation have lowered the costs of anti-access capabilities such that both states and nonstate actors can contest "America's 60-year-old dominance over the global commons and its ability to maintain their openness."[4]

As a result, new concepts such as "AirSea Battle" are being developed to "set the conditions at the operational level to sustain a stable, favorable conventional military balance throughout the Western Pacific region."[5] In general terms, AirSea Battle envisions integrated Air Force and Navy operational concepts that mitigate missile threats to U.S. bases; correct imbalances in strike capabilities between the United States and China's People's Liberation Army (PLA) in the Western Pacific; enhance undersea operations; offset the vulnerabilities of space-based command and control (C2) and intelligence, surveillance, and reconnaissance (ISR) systems; increase interoperability; and enhance electronic and cyber warfare capabilities.[6] It would do so by improving the "integration of air, land, naval, space, and cyberspace forces to provide

---

[1] U.S. Department of Defense, *Sustaining U.S. Global Leadership: Priorities for 21st Century Defense*, Washington, D.C., January 2012a, p. 2.

[2] President Obama emphasized that "while the U.S. military will continue to contribute to security globally, *we will of necessity rebalance toward the Asia-Pacific region*" (U.S. Department of Defense, 2012a, p. 2; emphasis in original).

[3] Andrew F. Krepinevich, *Why AirSea Battle?* Washington, D.C.: Center for Strategic and Budgetary Assessments, 2010, pp. vii–viii, 13. Krepinevich states, "China is engaged in a military modernization effort whose principal purpose appears to be to deny the United States the ability to sustain military forces in the Western Pacific" (p. 3).

[4] Abraham M. Denmark and James Mulvenon, eds., *Contested Commons: The Future of American Power in a Multipolar World*, Washington, D.C.: Center for a New American Security, January 2010, p. 6.

[5] Jan van Tol, Mark Gunzinger, Andrew F. Krepinevich, and Jim Thomas, *AirSea Battle: A Point-of-Departure Operational Concept*, Washington, D.C.: Center for Strategic and Budgetary Assessments, May 18, 2010, p. xi.

[6] van Tol et al., 2010, p. xiv.

combatant commanders the capabilities needed to deter and, if necessary, defeat an adversary employing sophisticated anti-access/area-denial capabilities."[7]

However, should the United States need to project power into China's near abroad, the technical and logistical challenges are daunting and will increase as China continues to invest in advanced anti-access and other systems and to develop complementary strategies. These capabilities are also incredibly expensive.[8] While it is important to develop concepts such as these, a complementary approach is to employ inexpensive anti-access technologies similar to those used by the PLA to significantly raise the cost of a conflict for China and, should deterrence fail, to drastically limit China's ability to inflict damage off the Asian mainland. Such approaches could be used in concert with an AirSea Battle approach or alone. This report explores such an approach: using ground-based anti-ship missiles (ASMs) as a tool around which to build a U.S. anti-access/area-denial (A2AD) strategy to counter China's maritime power projection capabilities and limit its maritime freedom of action. The primary goal of the analysis presented here is to demonstrate the tactical and operational viability of acquiring, transporting, and employing these systems.

Because this report focuses on operational and tactical issues, it is important to note that we make no attempt to comprehensively analyze the regional political, economic, and military dimensions of these operations; that is beyond the scope and purpose of this research. Should a conflict with China arise, the second- and subsequent-order effects would be complex, severe, and hard to forecast. As a result, much of this analysis assumed a wartime strategic context in which efforts to limit Chinese maritime freedom of action would be seriously considered and the stakes would be very high. This permits us to demonstrate the potential for a land-based ASM force to contribute to a "far blockade" while leaving the judgment of whether and when to use such a course of action to national-level decisionmakers. We also assumed that some or all of the regional states involved would be supportive of such a course of action in the context of a wider war. As such, the presentation focuses on operational and tactical issues, albeit ones with strategic implications.

We do, however, make one observation at the strategic level: The approach presented here is particularly appealing because it does not require the stationing or projecting of large-scale forces inside the first island chain, with all the implications that such a measure would have

---

[7] U.S. Department of Defense, *Joint Operational Access Concept*, version 1.0, Washington, D.C., January 17, 2012b.

[8] The approximate unit costs for some systems likely employed in the AirSea Battle operational concept are as follows, with amounts in fiscal year (FY) 2012 dollars (cost data from the Congressional Research Service, U.S. Government Accountability Office, and the U.S. Department of Defense Comptroller): the latest *Burke*-class Aegis destroyer ($2.03 billion); estimated *Zumwalt*-class destroyer ($3.09 billion); Littoral Combat Ship ($573.4 million); *Los Angeles*-class attack submarine ($1.58 billion); *Virginia*-class attack submarine ($2.79 billion); *Seawolf*-class attack submarine ($2.40 billion); B-2 Spirit ($1.63 billion); B-1B Lancer ($398.1 million); EA-18G Growler ($86.5 million); FA-18 Super Hornet ($80.1 million); F-22 Raptor ($179.7 million); and the estimated F-35 Joint Strike Fighter ($197.0 million). Prospective systems are also projected to be expensive. See, for example, Amy Butler, "Can USAF Buy a $550 Million Bomber?" *Aviation Week and Space Technology*, April 2, 2012.

Note, too, that these are just the acquisition costs of the individual systems; life-cycle costs include maintenance, fuel, personnel, and training and are significantly higher, as are the costs of deployment and employment (e.g., companion and support systems). By comparison, the transporter erector launcher portion of the BGM-109G ground-launched missile costs an estimated $4.19 million (2012 dollars; calculation based on the approach used by Richard Betts, *Cruise Missiles: Technology, Strategy, Politics*, Washington, D.C.: Brookings Institution Press, 1981, p. 104). In addition to a ground system, anti-ship missiles could be launched by many of the air and naval platforms above; unit costs for anti-ship cruise missiles range from $313,000 (for the UK's Sea Skua) to $1.2 million or more for the U.S. Harpoon or the latest-generation French Exocet.

for the Chinese and regional security situation. Indeed, the United States would not want to position these types of forces in the region unless tensions with China were very high. These modest-scale forces could be moved quickly into place if Chinese aggression indicated a need and if countries in the region were willing to accept them.[9] Furthermore, because China's projected economic and technical development over the coming decades will continue to tip the balance of military power in its favor in areas that it can influence from its territory, significant U.S. power projection and basing inside the first island chain may be increasingly challenging. Barring some currently unforeseen technical or other developments, fielding capabilities that do not present large or fixed targets would be useful.

The remainder of this report provides an overview of our approach and discusses why it makes sense for the United States and its allies in the region to consider an "anti-access" strategy.[10] It also briefly outlines how a ground-based ASM approach would need to fit into a joint approach to A2AD (it is not a stand-alone capability) and explores how these capabilities could combine with regional geography in a way the has a significant impact on Chinese strategic interests. Finally, the report briefly considers some logistical and procurement issues and concludes with final observations. Two appendixes supplement the findings presented here: Appendix A includes a list of selected ASM systems that are capable of being launched from the ground, and Appendix B offers a more complete accounting of our geospatial analysis of ASM capabilities in strategic waterways in the region.

Approach

As a means of illustrating the land-based ASM concept, we examine how they could be used to help establish a "far blockade" of China. We note that if the U.S. military had such capabilities, it could use them in a host of ways, ranging from security cooperation initiatives to help regional friends and allies establish their own anti-access capabilities to using them to interdict warships or (if supplemented by other assets) form a full blockade in times of war. We also recognize that the ability to implement such a blockade already exists in the U.S. Navy and Air Force. However, additional tools to implement it would provide U.S. commanders with more flexibility, would free up naval and air assets for other missions, and could prove less expensive than an air-naval-only approach.[11]

To determine what is possible and to illustrate potential capabilities, we conducted a missile-by-missile comparison of 45 current anti-ship cruise missiles.[12] These missiles are very popular with armies in the region; China, Indonesia, Malaysia, Vietnam, and Brunei are all believed to possess multiple types of ground-launched ASMs, for example (and some possess

---

[9] The agility with which they could be moved would depend on the force package that was deployed. For example, if they required a large ground security component and the attendant logistics capabilities, they would no longer be so agile. Keeping them small may be important for several reasons not examined here.

[10] *Anti-access* as used here implies denying China access to maritime areas in the vicinity of and outside of the first island chain.

[11] See footnote 8 on page 2 for comparative system costs. Full costing data are not available.

[12] This report focuses on ASMs that are designed to be launched from ground-based platforms, such as trucks or coastal batteries. Appendix A provides a partial listing of missiles that are available on the world market, along with information about their capabilities.

missiles that are fired from other platforms as well). A description of some of the missiles considered can be found in Appendix A.

We assessed the likely effectiveness of land-based ASMs by exploring the technical potential and impact of a U.S. anti-access strategy that could challenge Chinese maritime freedom of action. In our research approach, we assumed that the ability to cut off Chinese seaborne access beyond the first island chain would serve as a major deterrent and would, in the event of a conflict with China, have a significant effect on China's ability to attack its overseas neighbors and wage a prolonged war. Furthermore, this capability does not require the permanent stationing of assets in the Western Pacific, and, as such, it need not be presented as an effort to contain China. Rather, it should be seen as a capability that could be used if China initiated a conflict. Finally, although responsibility for developing and executing the U.S. military strategy in the Pacific will fall principally to the Office of the Secretary of Defense, the Joint Staff, U.S. Pacific Command (USPACOM), and the U.S. Navy, this report illustrates the potential for U.S. land forces to significantly contribute to USPACOM's efforts.

To illustrate the potential of land-based ASMs while acknowledging that a land-based-only approach would not be practical, we examined the possibility of cutting off Chinese sea routes using land-based ASMs only; we present a brief outline of the joint requirements for their implementation here. Not only would this have a significant effect on China's ability to project power, but it would also vastly expand the set of military problems for the PLA, should it consider initiating a conflict with its neighbors or U.S. partner nations. Specifically, because these missile systems are relatively easy to operate and are operationally and tactically mobile (i.e., they are not fixed targets), the PLA would have to search across a huge number of locations and have assets within range to interdict them. Furthermore, as the Israelis saw in their 2006 war in Lebanon, finding and destroying mobile missile launchers is extraordinarily difficult and requires a significant ability to mass intelligence assets and power at critical points.[13] Missile systems that could be placed in many locations over thousands of miles of island chains would significantly dilute the ability of PLA missile and air forces to be effective.

Finally, this report's purpose is to illustrate capabilities, not to make detailed recommendations about force structure or doctrine. Importantly, our intent is as much to encourage strategists to think of new approaches as it is to propose that the U.S. military consider developing a capability. This is but one such approach; there are others that also should be considered.

Contributions of a Land-Based ASM Approach to Coalition Efforts

China is a powerful nation with broad regional and global interests. It depends heavily on freedom of the seas for trade and, increasingly, to pursue its territorial and other interests. Furthermore, as its distant security interests continue to grow and its navy continues to modernize, its incentives to employ sea-based assets for such purposes as noncombatant evacuation operations (as it did recently in Libya in 2011), counterpiracy operations (as it currently is doing in the Gulf of Aden), and power projection are likely to increase. Furthermore, should tensions increase between the United States and its allies and China, Chinese naval assets could create conditions to hinder a U.S.-led coalition's naval freedom of action. Such a move would almost

[13] See, for example, David E. Johnson, *Hard Fighting: Israel in Lebanon and Gaza*, Santa Monica, Calif.: RAND Corporation, MG-1085-A/AF, 2011, pp. 83–85.

certainly involve PLA naval assets operating beyond the first island chain. As such, an important element of any U.S.-led coalition's strategy would be to limit Chinese naval freedom of action.

We build the case for the potential of land-based ASMs later in this report, but here we assert that, if they are effective, they would contribute to efforts to limit PLA naval freedom of action in two ways: by contributing to a "far blockade" centered on the first island chain and by increasing the PLA navy's risk when operating within firing range of coalition country shores. That said, limiting the PLA's ability to project naval surface power will, admittedly, not address all of its power-projection capabilities—specifically, the PLA's missile, air, and subsurface threats cannot be countered by ASMs. However, a land-based ASM component to joint and combined forces would indirectly contribute to efforts to counter the full spectrum of PLA assets. In particular, if land-based ASMs were used to help blockade China inside the first island chain, this would free up U.S. naval and air assets to pursue other missions in three important ways. First, freeing naval and air platforms from a blockade mission would allow USPACOM to dedicate more of these assets to the air and undersea domains. This could be critically important, for example, in protecting U.S. naval assets from PLA navy submarine threats, as well as in securing the very long sea lines of communication on which U.S. forces would rely. Second, more naval and air assets could facilitate broader geographic reach for U.S. joint forces or more intense efforts at critical points. And third, ASM assets would present a more robust "asymmetric" threat to Chinese naval surface groups, complicating their employment and forcing them to account for potential threats from multiple domains and axes that are difficult to identify.

Additionally, as discussed in more depth later in this report, land-based ASM assets would be mobile and need not operate from fixed locations. As such, if the PLA wanted to interdict them, it would have to devote a portion of its intelligence, targeting, shooting, and C2 assets to this task, significantly expanding its mission set, complicating its command decisionmaking process, and diluting its impact on other targets. Finally, such an expansion of U.S. and partner military assets and strategies may force the PLA navy to consider moving farther from the shoreline as it transits or patrols contested areas and, thus, into deeper waters where it would be more vulnerable to joint air, surface, and undersea assets.

A Joint Approach

A land-based ASM capability would be relatively easy to create in the U.S. armed forces, but it would be viable only as part of the joint force. In general, it would require access to other service (and perhaps national) sensor systems capable of identifying potential targets to engage, intelligence that can differentiate between commercial and combatant vessels, a C2 system that can receive and act on this information, and firing batteries that can respond to this C2 architecture and that—as in the case examined later—can range all passages through the straits that provide exits from the seas immediately surrounding China. Furthermore, because it is unlikely that the U.S. armed forces would fire missiles at civilian commercial shipping,[14] in a full blockade, assets that could challenge and, if necessary, board civilian ships trying to run a blockade would be required.[15] We provide a brief description of the elements of such a system, as well as a few of the major operational considerations that the U.S. armed forces would have to consider before discussing the potential for such an approach. While our examination of this approach demonstrates that these assets could be effectively employed, more analysis would be required to provide a complete operational approach.

An important element in an operational concept for employing ASMs is identifying potential targets. The United States has many assets that could be used for this purpose, including aerospace, ground, sea, and undersea assets with varying degrees of fidelity, sensor range, and communication capabilities. Some would be independent of the firing elements on the ground (e.g., satellites, aircraft, submarines), while others would be core elements of these systems (e.g., sensors on the missiles themselves) or otherwise could be employed by those operating the systems (e.g., unmanned aerial systems). We expect that these sensors will become more advanced and smaller over time, so deploying and employing them will become easier. In short, this architecture would need to include sensors that can identify targets at some distance and orient and queue firing units, at which point sensors on the systems would guide the missiles to the target. This would all be orchestrated by a C2 system supported by intelligence capabilities.

Once potential targets are identified, intelligence assets would need to determine whether they are actual targets. Considerations might include whether potential targets are war or merchant ships, their nationality, and with whom they have coordinated. Good answers to these questions will be critical to C2 centers that have to decide whether they should be engaged. (For example, under what conditions would the armed forces of a coalition fire on important neutral or even friendly vessels that were running a blockade?)

An effective C2 system will be critical to using—and avoid misusing—these systems. Such a system could be centralized or decentralized, depending on the doctrine and rules of engagement dictated by operational conditions. Decisions about what types of vessels to engage, and under what conditions, would have significant importance across a spectrum of

---

[14] For a host of political, diplomatic, legal, economic, environmental, and operational reasons, the National Command Authority would likely permit firing ASMs at neutral shipping only in very restricted circumstances. While blockade law is somewhat ambiguous and disputed, it is unlikely that firing on neutral commercial ships (e.g., cargo, tankers) would be considered legal, and doing so would have significant political ramifications. Modern commercial ships frequently have ownership, flagging, cargo, insurance, officers, and crew from multiple countries, further expanding the political ramifications. Environmental considerations would also be important. Given that modern oil tankers have roughly ten times the crude oil capacity of the Exxon *Valdez*, host governments are unlikely to allow such an environmental catastrophe off their shores in anything other than an existential threat, and they would be unlikely to allow the United States access and operational control if faced with this risk. Furthermore, while an ASM threat may deter commercial shipping and present new risks for Chinese planners, sinking a neutral civilian ship could substantially change U.S., host-country, and international public opinion in a conflict. Finally, actions to restrict naval freedom of action short of a blockade could also be envisioned. For example, the United States imposed a "quarantine" on Cuba during the October 1962 Cuban Missile Crisis to avoid some legal and political difficulties. For more details on the legal and political implications of blockades, see, for example, Lance Davis and Stanley Engerman, *Naval Blockades in Peace and War: An Economic History Since 1750*, New York: Cambridge University Press, 2006; International Committee of the Red Cross, *San Remo Manual on International Law Applicable to Armed Conflicts at Sea*, Geneva, Switzerland, 1994; and Matthew L. Tucker, "Mitigating Collateral Damage to the Natural Environment in Naval Warfare: An Examination of the Israeli Naval Blockade of 2006," *Naval Law Review*, Vol. 57, 2009.

[15] We note that the economic implications of such a blockade would be very significant. We do not examine these implications in this report but, rather, only note that ASMs alone—even if tied to an appropriate C2 network with adequate sensors—would not suffice to implement one without additional support.

concerns ranging from political to military to environmental. Different levels of command would make these decisions under different circumstances. In some cases, the President of the United States might be the appropriate decisionmaker; in others, the commander of the firing unit on the ground might be authorized to make that decision. Communication and network capabilities would have to be fielded to facilitate the sensor-to-decisionmaker-to-shooter connections. Appropriate targeting and decisionmaking rules would also be required, as would doctrine to guide the process.

In addition, the ability to deploy, employ, and provide adequate support for these systems would be required. Later in this report, we offer a quick analysis that approximates the airlift requirements for these systems, but these approximations will vary depending on which systems are fielded, their requirements for ammunition basic load and resupply, and the security and other support systems required to deploy with them (which would depend on the situation). These approximations are meant only to provide ballpark figures to approximate the lift requirements for the firing units alone; requirements for standard support and security packages are already known or can be easily approximated. Fast sealift is another option for moving ASM systems once in theater, and this would provide for less variance in lift capability based on logistical considerations. Additionally, host-nation systems could be part of the approach that we outline here, and may be preferable to U.S. units in many cases. On the one hand, they might require some resupply but not lift to operate, as they would already be in place, but on the other hand, they might provide their own security and supply, thus reducing U.S. logistical and lift requirements.

For these systems to be most effective, they would need to be mobile and have hiding locations that could be hardened. Having many potential firing positions across the vast expanse of the first island chain would significantly increase the targeting challenge for the PLA, as well as the firing units' chances of survival. Israel's challenges in finding and interdicting Hezbollah rockets in the 2006 war are one indication of the significance of this challenge. The extent to which these systems could have multiple hiding locations would depend on factors such as geography, infrastructure, and the support of the local population where they would be deployed. However, this comparison to firing rockets from southern Lebanon into Israel only goes so far. The number of locations from which these systems could fire would not be as numerous as those from which Katusha—and larger—rockets could be launched in southern Lebanon, because ASMs must be in close proximity to the sea. Finally, the firing teams would need caches for additional rounds that they could easily access, which would also require adequate geography, infrastructure, and local conditions.

Should a U.S. commander want to use these assets as part of a full blockade, the ability to challenge and board commercial ships could be provided by helicopters, small boats (if ranges to targets were not too great), or friendly navies and coast guards. We do not address these issues here, other than to note that such capabilities would be necessary in some circumstances and might be provided by the service that owned the firing units or by joint or coalition forces.

The remainder of this report assumes that these conditions are or can be put into place. It focuses primarily on the capabilities of the missile systems, themselves, and how they could be used in the context of the geography surrounding the first island chain.

Feasibility of a Land-Based ASM Blockade

Is it possible to deny China maritime freedom of action by blocking critical maritime chokepoints using land-based ASMs?[16] The appeal of such an approach is obvious; there are very capable missiles on the world market, and they provide a relatively inexpensive method to hold at-risk, expensive, hard-to-replace, and high-value assets, such as naval ships, from a long distance.

To illustrate what is possible using land-based ASMs, we examined how a distant blockade of chokepoints in the Asia-Pacific region could be built in phases.[17]

Strait of Malacca

The first phase we analyzed was blocking the Strait of Malacca by providing ASM coverage over the strait. Both Indonesia and Malaysia have robust arsenals of medium-range ASMs (U.S. forces currently have none). If both countries were willing to dedicate their missile capabilities to covering the strait or to permit other coalition partners to do so, the systems could effectively engage targets anywhere along approximately 730 km of the strait (see Figure 1).[18] Thus, even a relatively fast navy ship transiting the strait would be in the range of missiles for half a day.[19] Furthermore, such a capability would be hard for the PLA to defeat without put-

Figure 1
Ground-Launched Anti-Ship Missiles Currently in Indonesia and Malaysia

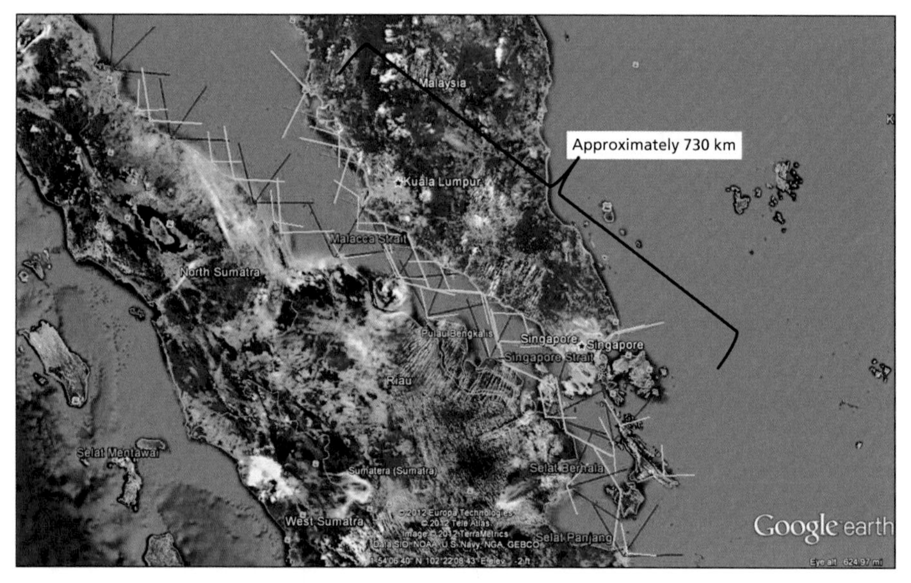

SOURCE: Google Earth, with overlays based on authors' geospatial analysis.
RAND *TR1321-1*

---

[16] It is useful to note that while this analysis focuses on the Western Pacific, a similar analysis could easily be done for other threats in other parts of the world.

[17] We derived missile capabilities, such as ranges, payload, and guidance systems, from open sources, which led to some discrepancies. Specific sources are identified later in this report.

[18] For detailed geospatial depictions of how anti-ship missiles could shut down all shipping routes to China, see Appendix B.

[19] Assuming that naval vessels could sail at 30 knots and merchant vessels at 15–20 knots.

ting land forces on the ground to locate these missile systems, as they are mobile and thus do not present fixed targets for missiles or air forces.

Longer-range ASMs would cause ships to be in range of missile batteries for even longer. These weapons are also in the arsenals of some regional countries. For example, Indonesia's C-802 anti-ship missile, a version of the Chinese YJ-2 with a range of at least 120 km, would make it the farthest-reaching ground-launched ASM in the region. However, Taiwan's Hsiung Feng III missile (130-km range) would extend the area subject to ASM fire to approximately 150 km to the northwest,[20] and Norway's Naval Strike Missile or Sweden's RBS-15 Mk III (both with advertised effective ranges of 200 km) could effectively cover more than 1,200 km of the Strait of Malacca (see Figure 2).[21] Finally, the BrahMos PJ-10, jointly developed and produced by India and Russia, has a range of approximately 300 km when launched from ground platforms. This range extends the complete coverage of the strait to approximately 1,500 km.[22] However, the BrahMos is twice as long and more than six times as heavy as the Naval Strike Missile, reducing its strategic and tactical mobility relative to other systems.

Figure 2
Ground-Launched Naval Strike Missile and RBS-15 Placed in Indonesia, Malaysia, and Thailand

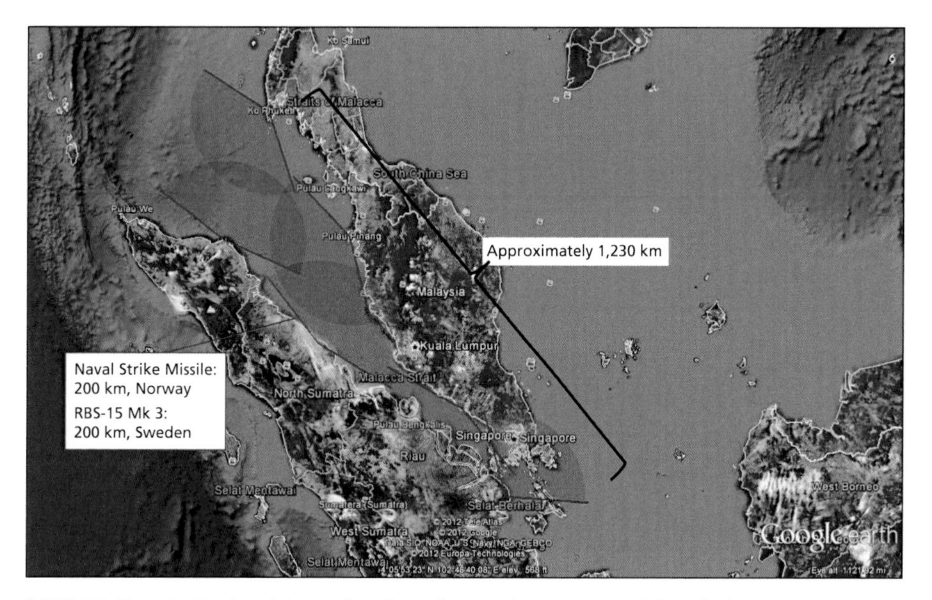

SOURCE: Google Earth, with overlays based on authors' geospatial analysis.
RAND *TR1321-2*

---

[20] The Hsiung Feng III also uses a ramjet motor that allows it to travel at approximately mach 2.0, which makes it one of the faster anti-ship missiles, giving surface ships less time to respond and counter attacks.

[21] The extended range, along with the Naval Strike Missile's capability to travel over ground and sea, would increase the survivability of missile systems by allowing them to be in more locations on the ground, making it even more difficult to locate them. The Naval Strike Missile is also the only fifth-generation ASM in production today. It boasts a number of advanced features that make it both more survivable (such as a low radar cross section and an extreme sea-skim flight mode) and more lethal (such as a high penetration capability, advanced guidance systems, and easier mission planning software) than other available ASMs.

[22] The BrahMos PJ-10 is also one of only a few supersonic cruise missiles. A second version allegedly being developed by India and Russia is likely to be even faster than the original.

This analysis of selected ASMs illustrates the degree to which specific capabilities can satisfy the requirements of a land-based maritime blockade. Such ASMs currently exist and can be easily purchased; developing a force structure to realize such capabilities does not require developing new weapon systems and could be implemented rapidly. Moreover, because these ASMs cost much less than the naval systems they target, they may prove to be cost-effective.

However, closing the Strait of Malacca does not amount to denying China maritime freedom of action. If the strait were shut down, a ship heading to or from the Indian Ocean could simply use the next-closest waterways, the Sunda and Lombok straits, or take even longer routes to get to China, thus evading the blockade. For example, a ship traveling from the Indian Ocean would have to sail an additional 550 nautical miles around the island of Sumatra using the Sunda Strait (the next-shortest route) to get to Shanghai.[23] Very large ships avoid this strait and prefer to use Lombok Strait, which extends a ship's transit by about 1,600 nautical miles (about three and a half days of travel time at a speed of 14–16 knots).[24]

Next, we illustrate how land-based ASMs could be used to interdict all sea lanes leading to China, thus making it, in effect, a landlocked nation.

Straits of Sunda and Lombok and the Java Sea Routes

The geographically narrow Sunda and Lombok straits can be easily covered with short-range missiles. These straits are significantly shorter than the Strait of Malacca, however, giving missile systems less space on either end to disperse and making them more vulnerable to enemy targeting. Because shipping could be rerouted around Lombok to the east and back toward the South China Sea to avoid these straits, it may also be valuable to use ASMs to separate the South China Sea from the Java Sea. Again, this can be done solely with anti-ship cruise missiles, given a range of no more than 100 km.[25] Figure 3 illustrates the situation after the second phase of an effort to limit Chinese maritime freedom of action, with the straits of Malacca, Sunda, and Lombok blocked.

If this second phase were successfully implemented, maritime shipping would be forced to either sail south of Australia or attempt to transit unblocked portions of the waterways surrounding South East Asia before accessing the Pacific Ocean and, ultimately, China. Regardless of which route they take, vessels would be forced to choose a course between the Philippines and Taiwan or between Taiwan and Japan.

Taiwan, Japan, and the Philippines

Should Taiwan and Japan be involved in a future scenario, ASM-based threats emanating from their territory would offer another capability to complicate PLA naval (PLAN) operations. For example, ground-launched ASMs located in Taiwan with a range of no more than 100 km, along with missiles with an effective range of 200 km positioned in Okinawa, could effectively deny all naval traffic south of Okinawa. Another possibility is to position missiles with a 200-km range solely on the Ryukyu Islands, which would also effectively close the area south

---

[23] Authors' calculations using Google Earth. The Sunda Strait is narrow, has a very shallow eastern entrance and strong tidal flows, and includes such obstacles as drilling platforms, a volcano, and small islands that make its transit very challenging.

[24] Lehman Brothers, *Global Oil Chokepoints*, January 18, 2008.

[25] Given that all these straits and naval passages lie within Indonesian influence, that country's support is critical to this operation.

Figure 3
Maritime Routes Left Open After Phase 2

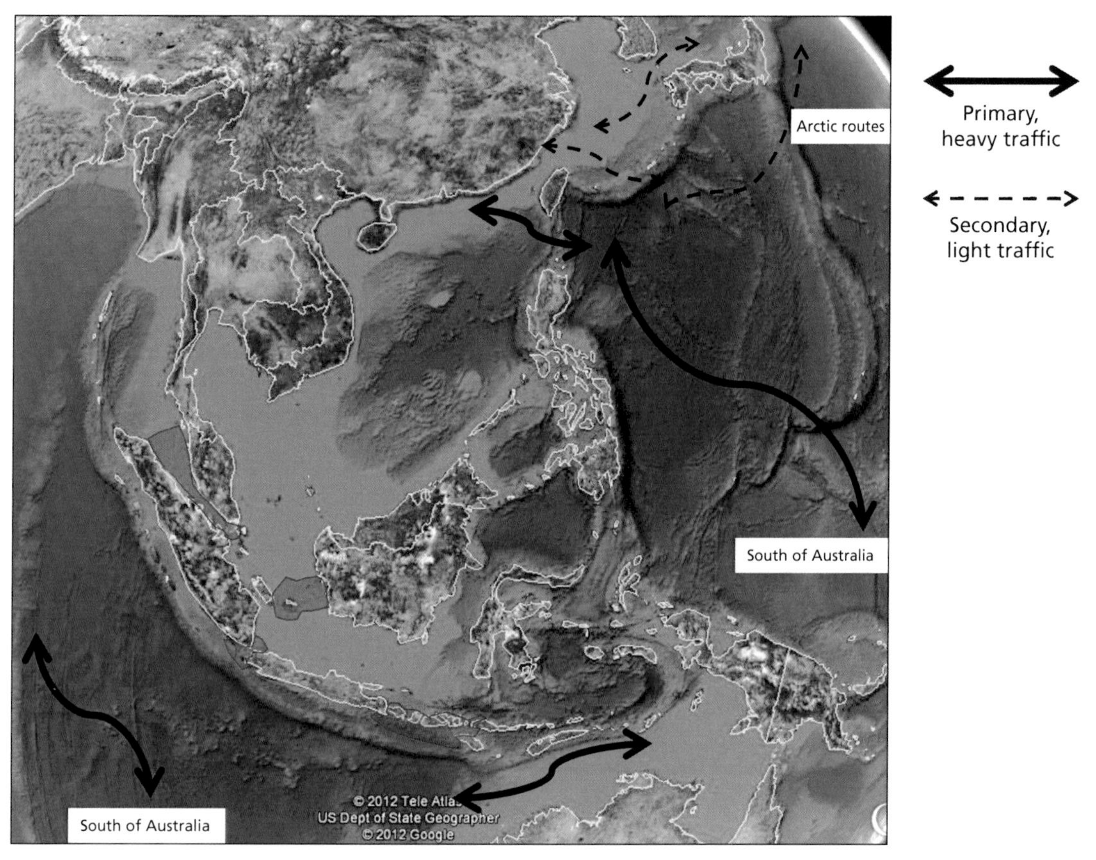

SOURCE: Google Earth, with overlays based on authors' geospatial analysis.
NOTE: Red areas are locations where access is denied by ASMs.
RAND *TR1321-3*

of Okinawa. The area between Okinawa and mainland Japan could be effectively covered by ASMs with a 100-km range in Japanese territory alone.

An implied task of this operational concept is to prevent Chinese forces from capturing these strategic islands. As part of a joint and combined effort, the very same ground-based A2AD systems used for the blockade would play a key role in these operations. In short, should China attempt to take these islands, it would incur asymmetric challenges and expenses similar to those the United States faces when considering how to project power inside the first island chain. However, China's capture of these islands would not necessarily open the waterways; rather, it would simply render infeasible a ground-based ASM means of closing them. Air and naval assets would still be available to keep these waterways closed, with ground-based systems to supplement them on the remaining waterways.

Similarly, the space between the Philippines and Taiwan, referred to as the Luzon Strait, and that between the Philippines and Borneo could be covered by 100-km-ranging missiles positioned in the Philippines, Taiwan, and Malaysia or by 200-km-ranging missiles (in the case of Taiwan) or shorter-range missiles (in the case of Borneo) fired solely from Philippine islands. The closure of these areas would significantly limit all naval activity, but more strategic depth could be achieved by also denying transit through the waters between Australia and

Indonesia. While such a move is not a necessity, the requirements would include the use of cruise missiles with a range of approximately 300 km (such as the BrahMos PJ-10) positioned in both Australian and Indonesian territory.

Figure 4 presents the maritime situation after this third phase of the operation is implemented.

Japan and South Korea

If other routes were contested, China could rely on arctic routes that transit between Japan and South Korea via the Korea Strait. To close these routes to and from the East China Sea, ASMs with a range of 200 km could be launched from either Japanese or South Korean territory (or 100-km-ranging missiles could be launched from both sides or from the Japanese island located near the middle of the strait).

As in the Strait of Malacca, however, operational flexibility and system survivability would increase with the use of both sides of this chokepoint. Once in place, the denial of

Figure 4
Naval Shipping Routes in Response to Phase 3 Area Denials

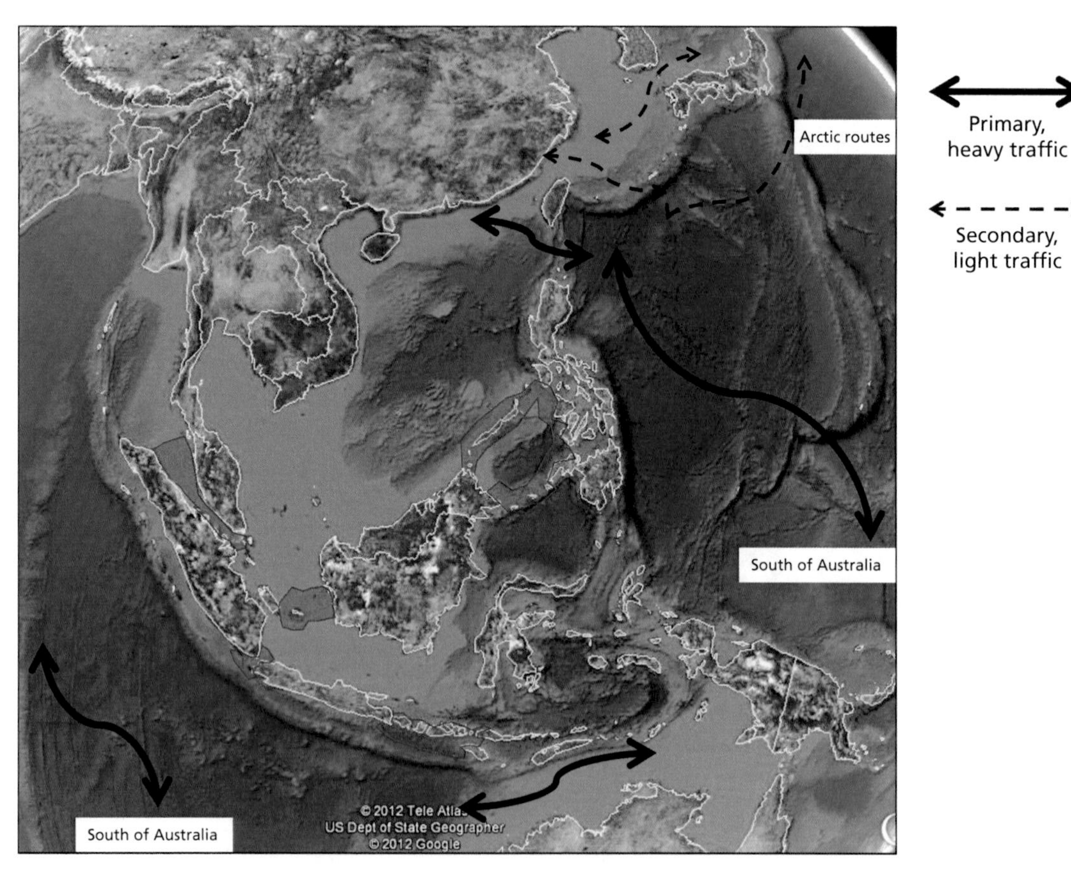

SOURCE: Google Earth, with overlays based on authors' geospatial analysis.
NOTE: Red areas are locations where access is denied by ASMs.
RAND *TR1321-4*

access from the East China Sea would "complete" a blockade of China using only ASMs, as depicted in Figure 5.[26]

The aggregate row in Table 1 identifies the requirements for this anti-access strategy. While it has been shown that the majority of maritime routes in the region can be covered by missiles with a range of no more than 100 km, both phase 3 options would require at least some missiles with an approximate range of 200 km. The aggregate row, then, identifies the minimum amount of foreign support required given the minimum technical capability of these 200-km missiles. To gain operational flexibility, support from all the countries discussed would be highly beneficial, and a number of advanced missile systems could provide technical advantages. However, it is feasible to deny maritime access to China with ground-launched

Figure 5
Areas Denied by Ground-Launched Anti-Ship Missiles in Partner Nations

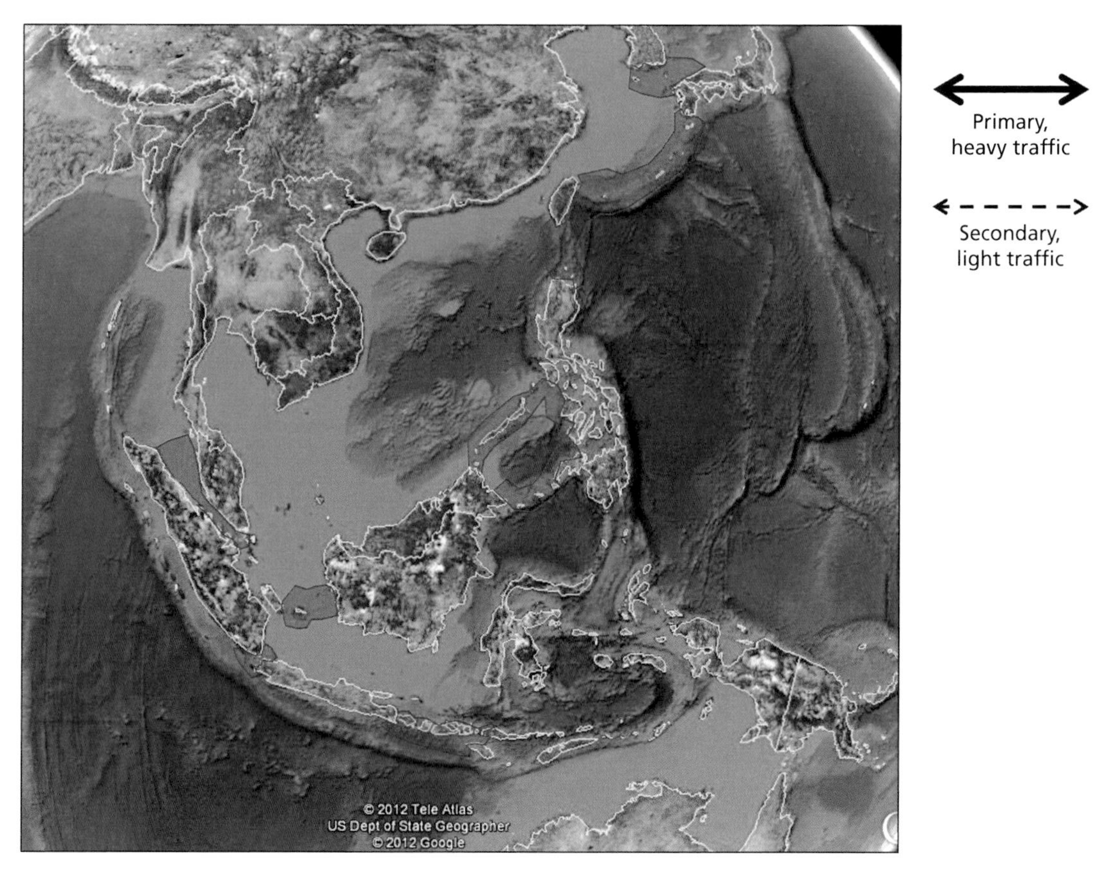

SOURCE: Google Earth, with overlays based on authors' geospatial analysis.
NOTE: Red areas are locations where access is denied by ASMs.
RAND *TR1321-5*

---

[26] The challenges of creating a complete blockade are documented in many sources. See, for example, Gabriel B. Collins and William S. Murray, "No Oil for the Lamps in China?" *Naval War College Review*, Vol. 61, No. 2, Spring 2008; and Bruce Blair, Chen Yali, and Eric Hagt, "The Oil Weapons: Myth of China's Vulnerability," *China Security*, No. 3, Summer 2006. We do not claim that ASMs alone would be successful where naval forces might not. We do note, however, that these systems would contribute to a USPACOM and coalition force effort to do so, should such an effort be necessary, freeing up naval and air assets for other missions.

Table 1
Phased Requirements to Deny Chinese Access to Regional Maritime Routes

Area	Minimum Missile Range Required (approx.)	Minimum Foreign Support Required	Optimal Foreign Support
Strait of Malacca	100 km (RBS-15 Mk2, ASM-2 type 96)	Indonesia or Malaysia	Indonesia, Malaysia, Thailand
Sunda Strait, Lombok Strait, and Java Sea	100 km (RBS-15 Mk2, ASM-2 type 96)	Indonesia	Indonesia
Ryukyu Island chain (2 options)	200 km (Naval Strike Missile, RBS-15 Mk 3)	Japan	Taiwan, Japan
	100 km (RBS-15 Mk2, ASM-2 type 96) in Taiwan and 200 km (Naval Strike Missile, RBS-15 Mk 3) in Japan	Taiwan, Japan	Taiwan, Japan
Luzon Strait (2 options)	200 km (Naval Strike Missile, RBS-15 Mk 3)	Philippines	Philippines, Taiwan
	100 km (RBS-15 Mk2, ASM-2 type 96)	Philippines, Taiwan	Philippines, Taiwan
Korean Strait (2 options)	100 km (RBS-15 Mk2, ASM-2 type 96)	Japan	Japan, South Korea
	200 km (Naval Strike Missile, RBS-15 Mk 3)	South Korea	Japan, South Korea
Combined requirements (all phases)	200 km (Naval Strike Missile, RBS-15 Mk 3)	Indonesia, Japan, Philippines	Indonesia, Malaysia, Thailand, Taiwan, Japan, Philippines, South Korea

SOURCE: Data from various open sources.

anti-ship cruise missiles with a range of no more than 200 km and employed from Indonesia, Japan, and the Philippines alone.

Logistical, Procurement, and Other Considerations

The approach illustrated here could be implemented using only currently available weapon systems and access to critical locations.[27] Next, we show that their deployment and employment could be quite flexible, permitting them to serve as a deterrent without requiring them to be permanently stationed in areas that the Chinese would see as threatening.

Logistical Considerations

Host-nation access is critical for employing an ASM blockade. However, prepositioning these systems in critical areas might be counterproductive, as Chinese officials could see it as threatening and demonstrating ill intent. Furthermore, even if the United States made the decision that it wanted to preposition such systems, it might not get access rights in circumstances less than a crisis. As such, if the United States had such systems, it would need to be able to rapidly move ASMs into the region from U.S. territory or from other prepositioned stocks.[28]

---

[27] This does not imply that the United States might not choose to develop systems that are tailored to its needs.

[28] Currently, the Army has prepositioned equipment in Southeast Asia, and the Marine Corps has prepositioning squadrons on Guam and Diego Garcia.

Many of these ASMs can be fired from a multitude of platforms and thus can be integrated with existing materiel and tactical requirements. For example, the RBS-15 Mk III is employed by the Finnish Navy using a Sisu SK242 truck, while the Croatian Army may use a Czechoslovakian-made Tatra truck to launch the same missile. Similarly, Poland is expected to deploy the Naval Strike Missile on the Jelcz truck, designed specifically for the Polish Navy.[29]

The ability to transfer these missile systems to multiple platforms and employ them from a number of vehicles with differing dimensions adds to their flexibility of use. However, this also makes it more difficult to determine specific lift and employment requirements. One method of generalizing these requirements to determine overall feasibility is to review analogous systems and current U.S. mobilization methods. Table 2 provides a rough comparison of the dimensions and weights of the ASMs considered here and the U.S. Patriot missile.[30] As Table 2 shows, the Patriot missile is longer, wider, and heavier than nearly all of the cruise missiles considered. With the exception of the BrahMos PJ-10, we can assume that the transportation requirements for each of these cruise missiles are no greater than that of the Patriot missile system, the requirements for which are well known.

The Army has determined the Patriot's minimum engagement package and identified a standard loading plan for C-5 and C-17 aircraft, so it can serve as a beginning point for planning estimates. The actual loads for any given mission would naturally be slightly different, as they would depend on mission-specific characteristics. The U.S. Army has established that the Patriot's minimum engagement package—which consists of two launchers, each with four missile canisters, eight total resupply missile canisters, and all the personnel and equipment needed to fully operate the system—could be delivered with the use of five C-5s or seven C-17s.[31] The BrahMos PJ-10 is a larger and much heavier missile than the Patriot and is

Table 2
Comparison of ASM Dimension and Weight Characteristics

Designation	Country of Origin	Length (m)	Diameter (m)	Launch Weight (kg)
Standard Patriot missile	United States	5.30	0.41	914
ASM-2 (Type 93, Type 96)	Japan	4.10	0.35	520
RBS-15	Sweden	4.35	0.50	790–805
Hsiung Feng III	Taiwan	5.10	0.38	660
Naval Strike Missile	Norway	3.96	0.70[a]	407
BrahMos PJ-10	India/Russia	8.20	0.67	3,000

SOURCE: Data from various open sources.

[a] Folded wingspan.

[29] "Purchase of New Anti-Ship Missiles in Poland to Build Sea Bases to Deal with the Russian Fleet," *Military of China, Force Comment Blog*, August 31, 2011.

[30] Although the Patriot system has a different purpose, its similarity in size and weight and the existence of analysis on its deployability makes this comparison useful for our analysis.

[31] Headquarters, U.S. Department of the Army, *Patriot Battalion and Battery Operations*, Washington, D.C., Field Manual 3-01.85, July 2010.

believed to use a Tatra 816 12×12 chassis for transportation and ground firing.[32] If the same (or a similar) type of vehicle were to be used as the ground platform for BrahMos PJ-10 missiles fired from Indonesia, Malaysia, and Thailand, it is likely that transportation requirements would increase should it be selected for U.S. forces.

Procurement Considerations

The global market for anti-ship cruise missiles is wide-reaching and complex. With dozens of missiles available from nearly as many manufacturers and countries, there are a number of avenues through which one may procure missiles. The missiles highlighted in this report were chosen because of their capabilities and assumed availability (e.g., not produced in China or Iran). The specific prices for each system, however, depend on the packages of systems purchased, specific requirements, and a number of other factors. The consultancy company Forecast International has conducted an in-depth analysis on the ASM market segment that includes ten-year detailed production forecasts, in-depth overviews of the principal market motivators and constraints, and calculations of projected manufacturer market shares by units and value.[33] However, for our purposes it is sufficient to note that missiles of this type are widely available for purchase and relatively inexpensive.[34]

Other Considerations

Other considerations will be important as the United States weighs the decision to add ASMs to its force structure and employ them. Here, we briefly touch on three.

Building Partner Capacity

If the United States created the force structure to enact an ASM blockade, it would provide not only a potent capability to combatant commanders but also opportunities to engage Asian partner nations in the region through security cooperation efforts that target this capability. Security cooperation is a mainstay of U.S. efforts to increase the capacity of partner nations, win access to territory, and influence other nations' behavior. Given the importance of ASMs in the first island chain, it is no surprise that many nations there have these systems. Whether they can employ them effectively, and whether they would do so as part of a coalition effort, are important questions. Yet, because the U.S. military does not have such systems, it has little capability to help build partner capacity in their use, so it may not be able to adequately influence the plans of allies and partners to deploy and employ them in concert with U.S. plans and efforts. As such, and in addition to operational reasons for having them, ground-based ASMs would expand the set of security cooperation options available to the United States.

Sabotaging Collision Hazard Infrastructure and Sea Mines

To increase the effectiveness of ASMs, the United States or its partners in the region could turn off, remove, or destroy current markers; sink ships; or emplace other obstacles to physically block part of a strait. This could be done in narrow and shallow areas (that may become

---

[32] "Brahmos (PJ-10)," *Jane's Strategic Weapon Systems*, April 25, 2012.

[33] Forecast International, *The Market for Anti-Ship Missiles, 2011–2020*, report and data package, Newtown, Conn., August 2011. The report and data package can be purchased through the company's website.

[34] As noted earlier, unit costs for anti-ship cruise missiles range from $313,000 (for the UK's Sea Skua) to $1.2 million or more for the U.S. Harpoon or the latest-generation French Exocet. This does not include launchers, radars, and other equipment.

targets for removal or destruction by the PLA as well). Effective obstacles could also shut down a strait to ships of a certain targeted size. A coalition could employ this tactic by itself or as part of a larger effort to canalize sea traffic into areas overseen by ASMs. Mines provide additional benefits; they can be low-tech and cheap to emplace. They also allow friendly ships to pass with guides, minimizing potential collateral damage to unintended targets. This tactic may require a large number of mines to be effective, and the mines must be monitored to prevent enemy removal.

Unmanned Vehicles

Unmanned underwater vehicles and aerial systems could also be used for ISR to assist in situational awareness, targeting, and anti-ship strikes. They can be man-portable and have the capability to remain in place for a long time. They are potentially important in extending the sensor networks that support U.S. forces and, in some cases, could provide the longer-range ISR capabilities that ASMs would need to operate beyond their line of sight. As the capabilities of unmanned systems increase, they could conceivably perform many of the tasks that this report posits for ASMs.

Defense Relations

The current U.S. strategy toward the Asia-Pacific region emphasizes the undeniable connection between Asian—including Chinese—and U.S. national interests. The United States seems to be pursuing a strategy to deepen and broaden existing alliances and partnerships with regional countries—particularly Japan, the Republic of Korea, Australia, Thailand, and the Philippines—while building new partnerships with other players, such as China, India, Singapore, Indonesia, Vietnam, Malaysia, and New Zealand.[35]

In November 2011, Secretary of State Hillary Clinton wrote an article for *Foreign Policy* titled "America's Pacific Century." In the article, Secretary Clinton suggests that the United States is well positioned to engage regional powers with economic and strategic partnerships, since it is "the only power with a network of strong alliances in the region, no territorial ambitions, and a long record of providing for the common good."[36] Pursuing the capabilities to create an ASM component of a USPACOM strategy and deploying those capabilities should therefore be balanced with how that would affect these relationships. As noted previously, the minimum and optimal foreign support for maritime denial operations would require the involvement of a few Asian countries, though deploying ASM systems permanently is neither required nor desirable at the present time.

Most of the nations upon which the United States would rely for access are strong partners or allies. However, Indonesia is arguably the most important for this strategy and has not traditionally been a U.S. partner (Malaysia, also not traditionally a U.S. partner, is important as well). Furthermore, while Indonesia currently accepts security assistance from the United

[35] Kurt M. Campbell, Assistant Secretary, Bureau of East Asian and Pacific Affairs, U.S. Department of State, *U.S.-Philippines Alliance: Deepening the Security and Trade Partnership*, testimony before the Committee on Foreign Affairs, Subcommittee on Terrorism, Nonproliferation, and Trade, U.S. House of Representatives, Washington, D.C., February 7, 2012.

[36] Hillary Clinton, "America's Pacific Century," *Foreign Policy*, November 2011.

States in the form of humanitarian assistance and disaster relief, peacekeeping operations, maritime security, and professionalization and reform,[37] it is also developing strong relations with China.[38] In recent years, Chinese and Indonesian militaries have interacted through defense consultation, exchange of visits, personnel training, equipment, joint training, maritime security, and multilateral security. Last year saw the first joint exercise of Chinese and Indonesian special forces, titled "Sharp Knife 2011," which enriched their capability to conduct joint actions.[39] On January 16, 2012, the Indonesian defense minister announced that Indonesia was entering a period of intense military expansion, and the Chinese defense minister met with the Indonesian ambassador to China to discuss increasing the two countries' military ties even further.[40] Then, in February 2012, the Indonesian defense minister visited Beijing to meet with Vice Premier Li Keqiang, who announced that China was ready to strengthen its bilateral relationship by expanding cooperation in the defense and security sectors.[41]

As a result, building partnerships that could lead to coalitions with such countries as Malaysia and Indonesia may be one of the biggest challenges to carrying out the strategy outlined in this analysis.[42] Yet, doing so is important for reasons far beyond the operational needs outlined in this report.

An Air-Sea-Land Concept

The Navy and Air Force currently possess the capacity to contest Chinese maritime freedom of action in Asia without land forces. However, doing so would require using expensive systems that would, if successfully targeted by Chinese forces, be difficult to replace. An inexpensive truck-mounted missile launcher in a Philippine jungle is considerably more difficult to locate and attack than an expensive naval warship patrolling the approaches to the Strait of Malacca—and yet both could contribute to blockade objectives. Furthermore, the demand for naval assets to control the sea lines of communication to U.S. bases in the Western Pacific and perform other missions in times of conflict would be significant. Land-based ASMs could help relieve some of these demands on the Navy (and Air Force). Additionally, positioning of ASM

---

[37] After a proposal in 2008 by Indonesian President Yudhoyono, the United States and Indonesia began a comprehensive partnership program in November 2010 to enhance bilateral interactions. This long-term pledge was reaffirmed by Presidents Obama and Yudhoyono one year later and consists of the three separate pillars: politics and security, economics and development, and sociocultural issues, education, science, and technology cooperation. See U.S. Agency for International Development, Indonesia, "Comprehensive Partnership," web page, last updated September 26, 2012; Office of the White House Press Secretary, "United States–Indonesia Comprehensive Partnership," fact sheet, Washington, D.C., November 19, 2011; "The Happening Place," *The Economist*, November 12, 2011; and U.S. Embassy in Jakarta, Indonesia, "American and Indonesian Defense Officials Participate in Defense Cooperation Seminar," February 15, 2012.

[38] Xinhua, "China, Indonesia Eye for Closer Military Links," *China.org.cn*, January 16, 2012a.

[39] "The Happening Place," 2011; Xinhua, "China-Indonesia Joint Training for Special Forces Ends," *China.org.cn*, June 17, 2011. The United States has also begun training the controversial Indonesian special forces, which have been accused of human rights abuses in the past.

[40] Ashley Woermann, "Indonesian Military Expansion Strengthens China Partnership," *Future Directions International*, January 25, 2012.

[41] Xinhua, "Chinese Vice Premier Stresses Strategic Partnership with Indonesia," *People's Daily Online*, February 22, 2012.

[42] In this regard, it is important to emphasize the significant risks to countries in the region that agree to cooperate with the United States.

systems throughout the first island chain would very significantly increase the PLA's targeting requirements, stressing its C2 systems and causing it to spread valuable intelligence, targeting, and attack assets over many possible firing positions across an arch of islands that is thousands of miles long rather than focusing on a few well-defined targets that contain the major U.S. bases in the region. Arguably, this would significantly decrease the relative effectiveness of PLA anti-access assets and increase the relative effectiveness of other U.S. and coalition efforts.

The current AirSea Battle concept understandably places a significant emphasis on the Navy and Air Force's capability to counter foreign A2AD threats. This report illustrates that creating an ASM capability in the U.S. ground forces could significantly dilute the A2AD threat and present a corresponding U.S. capability to an aggressor state that sought to project power over water. In short, developing and employing ASMs in the force structure of either the Army or Marine Corps would provide capabilities that could have a strategic effect.

Additionally, land-based ASMs would provide future U.S. presidents with capabilities that could extend a conflict without escalating it (e.g., by attack targets in China). This time could create the space for political solutions and de-escalate the conflict.

Finally, capabilities such as those presented here will become increasingly accessible to nations and, perhaps, nonstate actors. Armed, unmanned systems (aerial and under water) could have similar effects to ASMs. Keeping these capabilities out of the hands of rogue actors will likely be an important task—one that could be used to build ties with China in the form of nonproliferation regimes, because both China and the United States would have a large stake in such efforts.

Conclusions

Land-based ASMs are readily available on the world's arms markets, are inexpensive, and would provide significant additional capabilities to the United States if integrated into the Army or Marine Corps force structure. Their employment would require coalition and joint concepts and approaches, as well as support from joint assets, such as sensors, intelligence, and C2 systems. But the capabilities they could provide a coalition force would free up the Navy and Air Force for missions other than controlling maritime traffic (military or commercial) near land chokepoints. These capabilities would also significantly complicate the PLA's C2, intelligence, and targeting requirements and would raise the cost of a conflict for China (and other nations that depend on maritime freedom of action). While a detailed analysis of fielding these systems was beyond the scope of this report, we believe that having them in the inventory would further U.S. efforts to provide security cooperation assistance to partner nations, could help deter conflict, and could contribute to victory in a future conflict by increasing flexibility and expanding the set of tools available to U.S. commanders to implement plans. It is also likely that fielding ASMs would be cheaper—perhaps significantly so—than other means of deterrence.

APPENDIX A

Selected Anti-Ship Missiles Capable of Being Launched from Ground-Based Platforms

This appendix provides additional background on some of the ASM systems discussed in this report that can be launched from ground-based platforms. Table A.1 lists the characteristics of these systems, including their range, guidance systems, and countries of origin and export. It also lists the full launch capabilities of each system.

Table A.1
Selected Anti-Ship Missiles Capable of Being Launched from Ground-Based Platforms

Designation	Country of Origin	Range (km)	Guidance	Exported to	Launch Platform
MM-38 Exocet	France	40	INS, active radar	Cameroon, Chile, Colombia, Cyprus, Ecuador, Germany, Greece, Indonesia, Iraq, Ivory Coast, South Korea, Kuwait, Malaysia, Morocco, Nigeria, Oman, Pakistan, Peru, Qatar, Saudi Arabia, Taiwan, Thailand, Tunisia	Ship and ground
MM-40 Exocet	France	70	INS, active radar	Belgium, Brazil, Brunei, Cameroon, Chile, Colombia, Cyprus, Ecuador, Germany, Greece, Indonesia, Iraq, Ivory Coast, South Korea, Kuwait, Malaysia, Morocco, Nigeria, Oman, Pakistan, Peru, Qatar, Saudi Arabia, Taiwan	Ship and ground
BrahMos PJ-10	India/ Russia	300 or 500	INS, GPS, active and passive radar	Expected to be in South Africa, Chile, Brazil, and a host of countries in the Middle East and Africa	Ship, air, ground, and submarine
Otomat/Teseo	Italy	60–180	INS, datalink, active radar	Bangladesh, Egypt, Iraq, Kenya, Libya, Malaysia, Nigeria, Peru, Saudi Arabia, Venezuela	Ship and ground
ASM-2 (Type 93, Type 96)	Japan	100	INS, datalink, IIR		Air and ground
YJ-2/Eagle Strike/CSS-N-8 Saccade/C-802	China	120	INS, active radar	Iran, Pakistan	Ship, air, ground, and submarine
SSC-3 Styx	Russia	80	Autopilot, radio altimeter, active radar/IIR	Croatia, Cuba, Egypt, Ethiopia, Finland, Georgia, Germany, India, Indonesia, Iran, Iraq, North Korea, Libya, Poland, Romania, Serbia, Somalia, Syria, Tunisia, Ukraine, Vietnam, Yemen	Ground
RBS-15	Sweden	100–200	INS, radio altimeter, active radar	Croatia, Finland, Germany, Montenegro, Poland, Serbia	Ship, air, and ground
Hsiung Feng 3/ HF-3/Male Bee 3	Taiwan	130	Inertial, active radar with infrared seeker		Ship and ground
Naval Strike Missile	Norway	3–200		Poland	Ship and ground

NOTE: INS = inertial navigation system. IIR = imaging infrared. GPS = Global Positioning System.

APPENDIX B

Geospatial Analysis of ASM Capabilities in Strategic Asian Waterways

This appendix presents detailed geospatial depictions of how ASMs could shut down all shipping routes to China. Figures B.1–B.11 rely on aerial imagery from Google Earth with overlays based on the authors' geospatial analysis.

Figure B.1
Ground-Launched RBS-15 Mk 2 and ASM-2 (Type 96) Systems Positioned in Indonesia to Form a Blockade of the Sunda Strait

RAND *TR1321-B.1*

Figure B.2
Ground-Launched RBS-15 Mk 2 and ASM-2 (Type 96) Systems Positioned in Indonesia to Form Blockade of the Lombok Strait and Surrounding Passageways

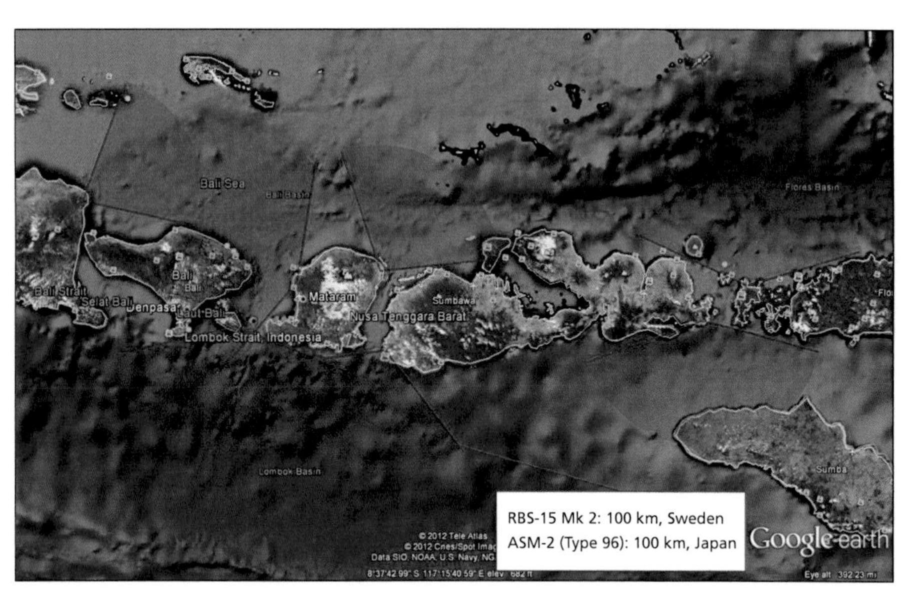

RAND *TR1321-B.2*

Figure B.3
Ground-Launched RBS-15 Mk 2 and ASM-2 (Type 96) Systems Positioned in Indonesia to Separate the Java Sea from the South China Sea

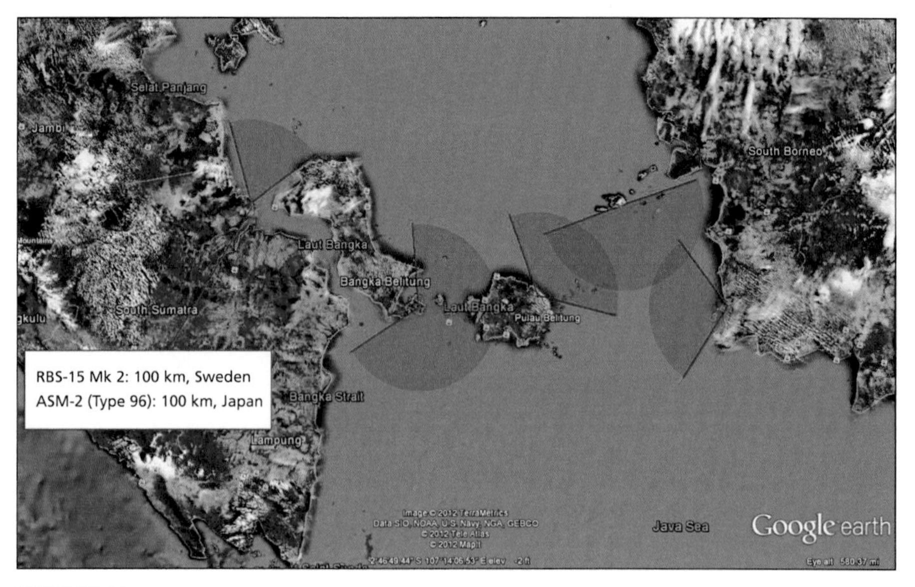

RAND *TR1321-B.3*

Figure B.4
Ground-Launched RBS-15 Mk 2, ASM-2 (Type 96), Naval Strike Missile, and RBS-15 Mk 3 Systems Positioned in Japan and Taiwan (South of Okinawa)

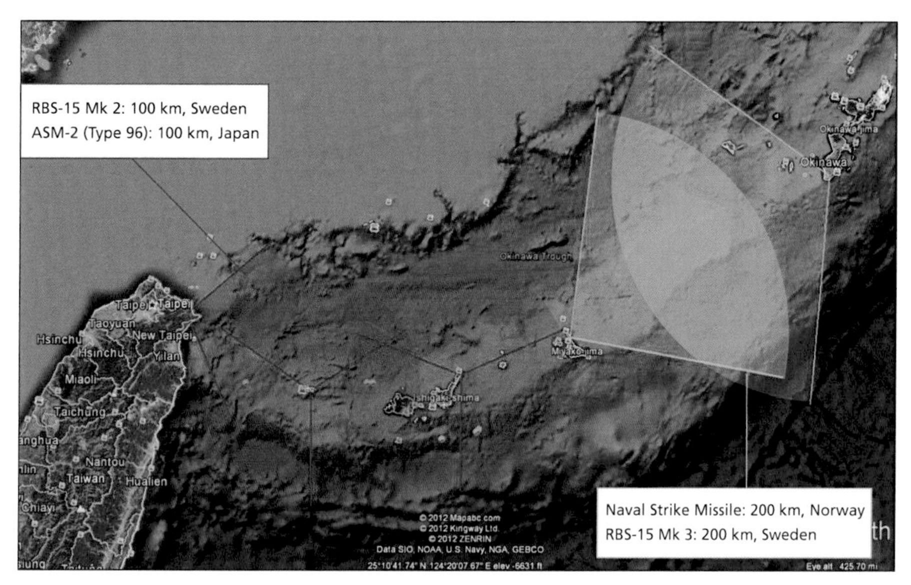

RAND *TR1321-B.4*

Figure B.5
Ground-Launched Naval Strike Missile and RBS-15 Mk 3 Systems Positioned in Japan (South of Okinawa)

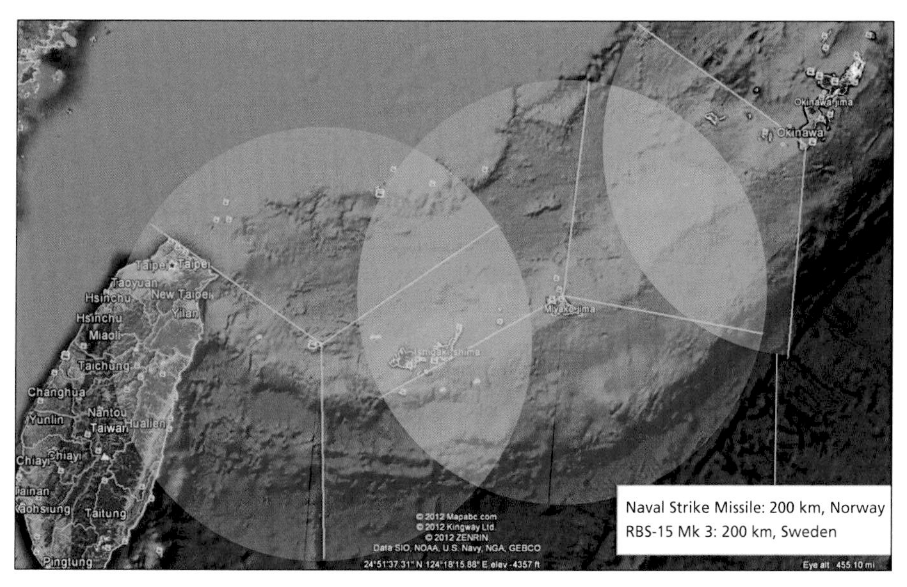

RAND *TR1321-B.5*

Figure B.6
Ground-Launched RBS-15 Mk 2 and ASM-2 (Type 96) Systems Positioned in Japan (North of Okinawa)

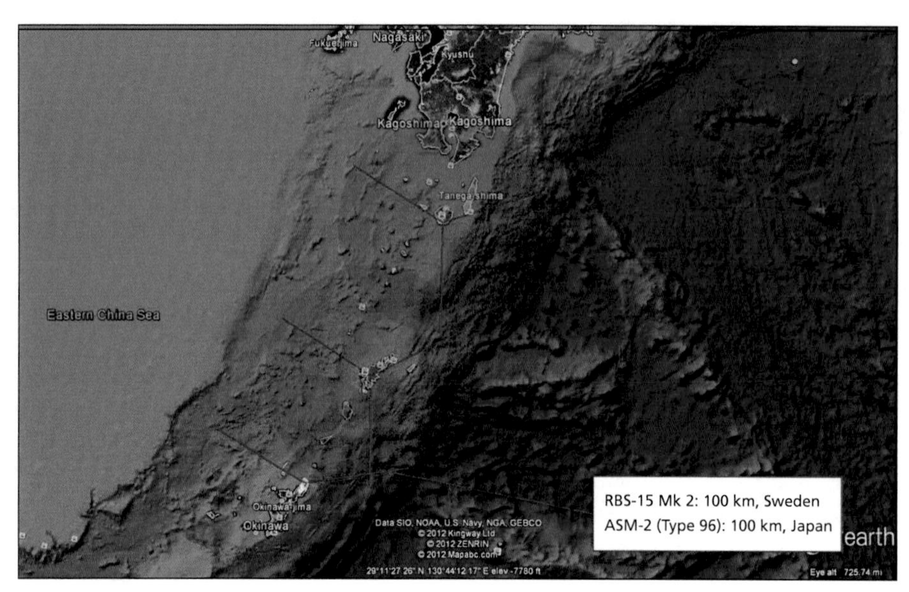

RAND *TR1321-B.6*

Figure B.7
Ground-Launched RBS-15 Mk 2 and ASM-2 (Type 96) Systems Positioned in the Philippines and Taiwan

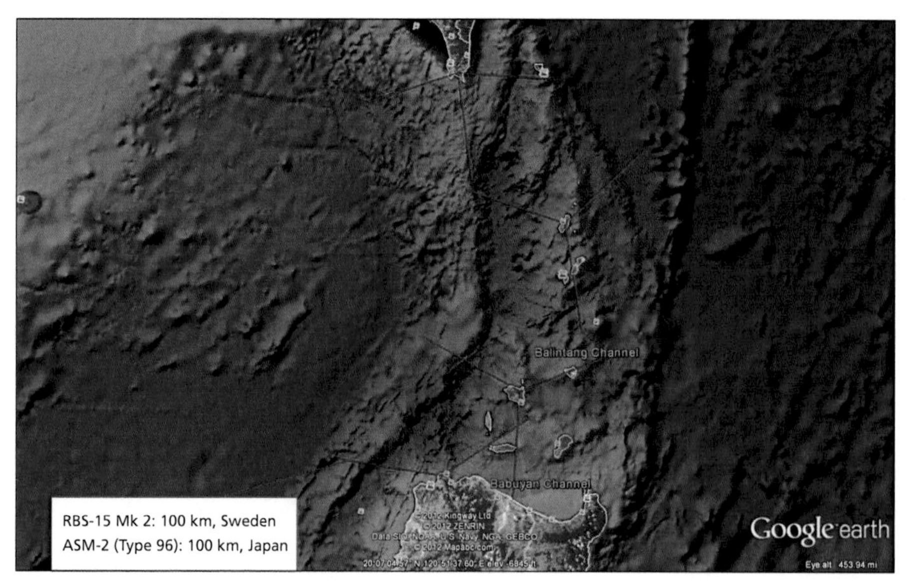

RAND *TR1321-B.7*

Figure B.8
Ground-Launched Naval Strike Missile and RBS-15 Mk 3 Systems Positioned in the Philippines

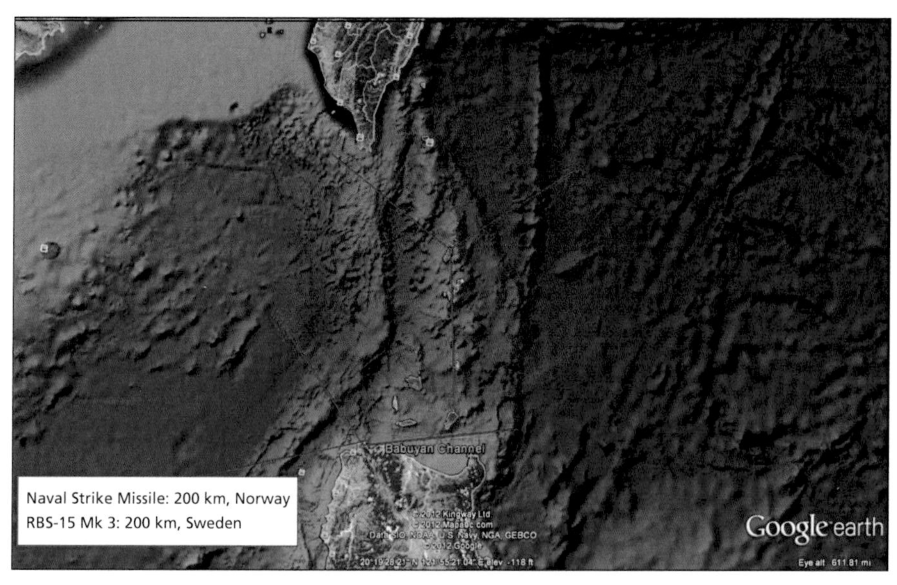

RAND *TR1321-B.8*

Figure B.9
Ground-Launched BrahMos PJ-10 Systems Positioned in Indonesia and Australia to Close the Savu and Timor Seas

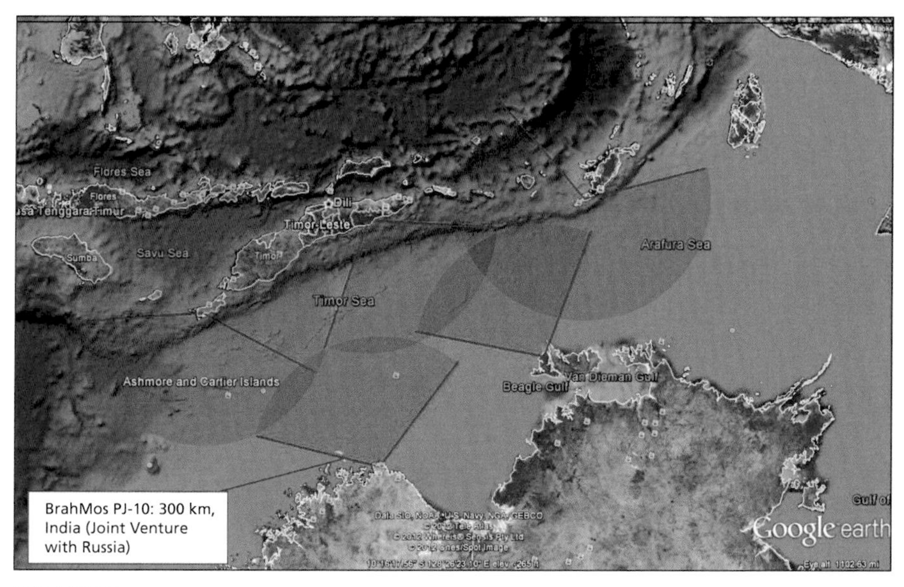

RAND *TR1321-B.9*

Figure B.10
Ground-Launched RBS-15 Mk 2 and ASM-2 (Type 96) Systems Positioned in Japan

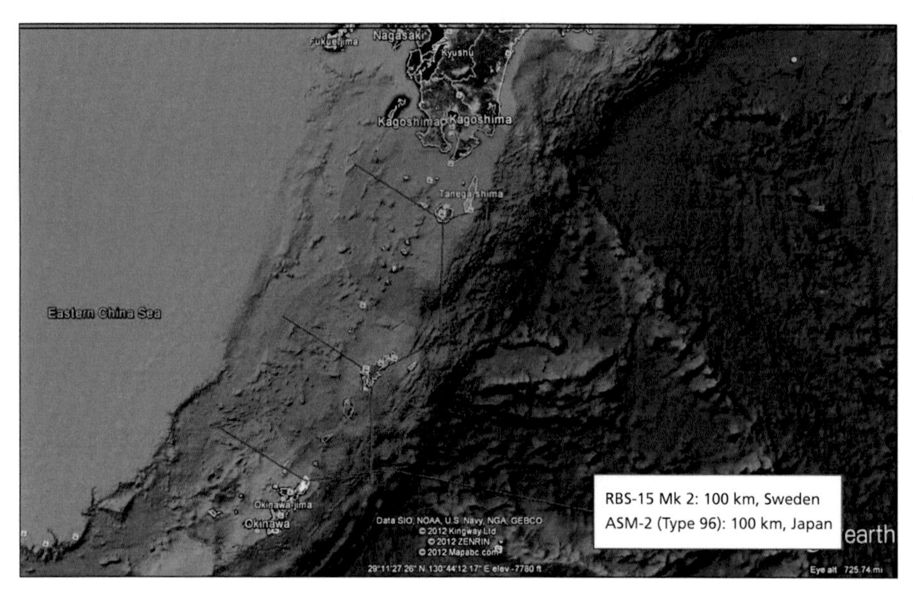

RAND *TR1321-B.10*

Figure B.11
Ground-Launched Naval Strike Missile and RBS-15 Mk 3 Systems Positioned in South Korea

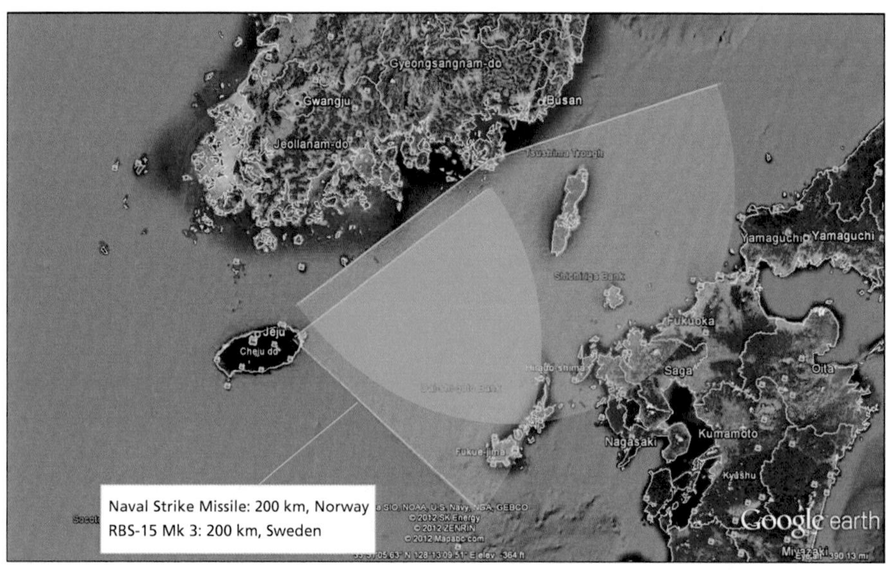

RAND *TR1321-B.11*

Bibliography

Alessi, Christopher, and Stephanie Hanson, *Expanding China-Africa Oil Ties*, Washington, D.C.: Council on Foreign Relations, February 8, 2012. As of October 1, 2012: http://www.cfr.org/china/expanding-china-africa-oil-ties/p9557

"Arctic Sea Routes Opening Up for China," *China Briefing*, July 21, 2011. As of October 1, 2012: http://www.china-briefing.com/news/2011/07/21/arctic-sea-routes-opening-up-for-china.html

Bateman, Sam, Catherine Zara Raymond, and Joshua Ho, *Safety and Security in the Malacca and Singapore Straits: An Agenda for Action*, Singapore: Institute of Defence and Strategic Studies, May 2006.

Berger, Brett, and Robert F. Martin, *The Growth of Chinese Exports: An Examination of the Detailed Trade Data*, Washington, D.C.: Board of Governors of the U.S. Federal Reserve System, International Finance Discussion Paper 1033, November 2011. As of October 1, 2012: http://www.federalreserve.gov/pubs/ifdp/2011/1033/ifdp1033.pdf

Betts, Richard, *Cruise Missiles: Technology, Strategy, Politics*, Washington, D.C.: Brookings Institution Press, 1981.

Blair, Bruce, Chen Yali, and Eric Hagt, "The Oil Weapons: Myth of China's Vulnerability," *China Security*, No. 3, Summer 2006, pp. 32–63.

Blair, Dennis, and Kenneth Lieberthal, "Smooth Sailing: The World's Shipping Lanes Are Safe," *Foreign Affairs*, May–June 2007.

Bradsher, Keith, "China Leading Global Race to Make Clean Energy," *New York Times*, January 30, 2010. As of October 1, 2012: http://www.nytimes.com/2010/01/31/business/energy-environment/31renew.html

"Brahmos (PJ-10)," *Jane's Strategic Weapon Systems*, April 25, 2012.

Butler, Amy, "Can USAF Buy a $550 Million Bomber?" *Aviation Week and Space Technology*, April 2, 2012.

Campbell, Kurt M., Assistant Secretary, Bureau of East Asian and Pacific Affairs, U.S. Department of State, *U.S.-Philippines Alliance: Deepening the Security and Trade Partnership*, testimony before the Committee on Foreign Affairs, Subcommittee on Terrorism, Nonproliferation, and Trade, U.S. House of Representatives, Washington, D.C., February 7, 2012. As of October 1, 2012: http://www.state.gov/p/eap/rls/rm/2012/02/183494.htm

Central Intelligence Agency, "East and Southeast Asia: China," *The World Factbook*, last updated September 11, 2012. As of October 1, 2012: https://www.cia.gov/library/publications/the-world-factbook/geos/ch.html

Clinton, Hillary, "America's Pacific Century," *Foreign Policy*, November 2011. As of October 1, 2012: http://www.foreignpolicy.com/articles/2011/10/11/americas_pacific_century

Collins, Gabriel B., and William S. Murray, "No Oil for the Lamps in China?" *Naval War College Review*, Vol. 61, No. 2, Spring 2008, pp. 79–95.

Corben, Ron, "Burmese Pipeline to China Under Construction, Despite Criticism," *Voice of America*, September 6, 2011. As of October 1, 2012: http://www.voanews.com/english/news/asia/Burmese-Pipeline-to-China-Under-Construction-Despite-Criticism-129369198.html

Davis, Lance, and Stanley Engerman, *Naval Blockades in Peace and War: An Economic History Since 1750*, New York: Cambridge University Press, 2006.

Denmark, Abraham M., and James Mulvenon, eds., *Contested Commons: The Future of American Power in a Multipolar World*, Washington, D.C.: Center for a New American Security, January 2010. As of October 1, 2012:
http://www.cnas.org/node/4012

Dieles, Nele, "A Silk Road for the 21st Century: Freight Rail Linking China and Germany Officially Begins Operations," *Shanghaiist*, July 4, 2011. As of October 1, 2012:
http://shanghaiist.com/2011/07/04/a_cargo_train_filled_with.php

Downs, Erica Strecker, *China's Quest for Energy Security*, Santa Monica, Calif.: RAND Corporation, MR-1244-AF, 2000. As of October 1, 2012:
http://www.rand.org/pubs/monograph_reports/MR1244.html

Forecast International, *The Market for Anti-Ship Missiles, 2011–2020*, report and data package, Newtown, Conn., August 2011.

Global Security.org, "South China Sea Oil Shipping Lanes," web page, last updated July 11, 2011. As of October 1, 2012:
http://www.globalsecurity.org/military/world/war/spratly-ship.htm

"The Happening Place," *The Economist*, November 12, 2011. As of October 1, 2012:
http://www.economist.com/node/21538218

Hays, Jeffrey, "Chinese Trade: World Economy, Container Ships and the WTO," *Facts and Details*, 2008. As of October 1, 2012:
http://factsanddetails.com/china.php?itemid=350&catid=9&subcatid=62

Headquarters, U.S. Department of the Army, *Patriot Battalion and Battery Operations*, Washington, D.C., Field Manual 3-01.85, July 2010.

International Committee of the Red Cross, *San Remo Manual on International Law Applicable to Armed Conflicts at Sea*, Geneva, Switzerland, 1994. As of February 12, 2013:
http://www.icrc.org/eng/resources/documents/misc/57jmsu.htm

Johnson, David E., *Hard Fighting: Israel in Lebanon and Gaza*, Santa Monica, Calif.: RAND Corporation, MG-1085-A/AF, 2011. As of October 1, 2012:
http://www.rand.org/pubs/monographs/MG1085.html

Kamrany, Nake M., and George Milanovic, "China's Growing Economic Strength in the 21st Century," *Huffington Post*, November 17, 2011. As of October 1, 2012:
http://www.huffingtonpost.com/nake-m-kamrany/chinas-growing-economic-s_b_1100416.html

Kazianis, Harry, "Anti-Access Goes Global," *TheDiplomat.com, Flashpoints Blog*, December 2, 2011. As of October 1, 2012:
http://the-diplomat.com/flashpoints-blog/2011/12/02/anti-access-goes-global

Krepinevich, Andrew F., *Why AirSea Battle?* Washington, D.C.: Center for Strategic and Budgetary Assessments, 2010.

Lehman Brothers, *Global Oil Chokepoints*, January 18, 2008.

Luft, Gal, "Fueling the Dragon: China's Race into the Oil Market," Potomac, Md.: Institute for the Analysis of Global Security, 2004. As of October 1, 2012:
http://www.iags.org/china.htm

Mills, Richard, "The New Steel Silk Road," *International Business Times*, November 2, 2011. As of October 1, 2012:
http://au.ibtimes.com/articles/241639/20111102/the-new-steel-silk-road.htm

Moe, Wai, "Burma-China Pipeline Work to Start in September," *Irrawaddy*, June 16, 2009. As of October 1, 2012:
http://www2.irrawaddy.org/article.php?art_id=16024

N'Diaye, Papa, and Nathan Porter, "China's Evolving Role in Global Trade," in *People's Republic of China 2011 Spillover Report—Selected Issues, a component report of People's Republic of China: Spillover Report for the 2011 Article IV Consultation and Selected Issues*, Washington, D.C.: International Monetary Fund, Country Report No. 11/193, June 28, 2011. As of October 1, 2012:
http://www.imf.org/external/pubs/ft/scr/2011/cr11193.pdf

Office of the White House Press Secretary, "United States–Indonesia Comprehensive Partnership," fact sheet, Washington, D.C., November 19, 2011. As of October 1, 2012:
http://www.whitehouse.gov/the-press-office/2011/11/18/
fact-sheet-united-states-indonesia-comprehensive-partnership

"Old King Coal," *The Economist*, February 23, 2012. As of October 1, 2012:
http://www.economist.com/node/21548237

"Purchase of New Anti-Ship Missiles in Poland to Build Sea Bases to Deal with the Russian Fleet," *Military of China, Force Comment Blog*, August 31, 2011. As of October 1, 2012:
http://www.9abc.net/index.php/archives/21658

Ramos, Fidel V., "Geopolitical Shifts in Global Power: The Rise of Asia," *Manila Bulletin*, October 9, 2010. As of October 1, 2012:
http://www.mb.com.ph/articles/281356/geopolitical-shifts-global-power-the-rise-asia

Rodrigue, Jean-Paul, Theo Notteboom, and Brian Slack, "Maritime Transportation," in Jean-Paul Rodrigue, Claude Comtois, and Brian Slack, eds., *The Geography of Transport Systems*, 2nd ed., New York: Routledge, 2012. As of October 1, 2012:
http://people.hofstra.edu/geotrans/eng/ch3en/conc3en/ch3c4en.html

Storey, Ian, "China's 'Malacca Dilemma,'" *China Brief* (Jamestown Foundation), Vol. 6, No. 8, 2006. As of October 1, 2012:
http://www.jamestown.org/programs/chinabrief/
single/?tx_ttnews[tt_news]=3943&tx_ttnews[backPid]=196&no_cache=1

Tucker, Matthew L., "Mitigating Collateral Damage to the Natural Environment in Naval Warfare: An Examination of the Israeli Naval Blockade of 2006," *Naval Law Review*, Vol. 57, 2009, pp. 161–202.

U.S. Agency for International Development, Indonesia, "Comprehensive Partnership," web page, last updated September 26, 2012. As of October 1, 2012:
http://indonesia.usaid.gov/en/about/comprehensive_partnership

U.S. Department of Defense, *Sustaining U.S. Global Leadership: Priorities for 21st Century Defense*, Washington, D.C., January 2012a. As of October 1, 2012:
http://www.defense.gov/news/Defense_Strategic_Guidance.pdf

———, *Joint Operational Access Concept*, version 1.0, Washington, D.C., January 17, 2012b. As of October 1, 2012:
http://www.defense.gov/pubs/pdfs/JOAC_Jan%202012_Signed.pdf

U.S. Embassy in Jakarta, Indonesia, "American and Indonesian Defense Officials Participate in Defense Cooperation Seminar," February 15, 2012. As of October 1, 2012:
http://jakarta.usembassy.gov/news/embnews_02152012.html

U.S. Energy Information Administration, *World Oil Transit Chokepoints*, August 22, 2012. As of October 1, 2012:
http://www.eia.gov/countries/analysisbriefs/World_Oil_Transit_Chokepoints/wotc.pdf

van Tol, Jan, Mark Gunzinger, Andrew F. Krepinevich, and Jim Thomas, *AirSea Battle: A Point-of-Departure Operational Concept*, Washington, D.C.: Center for Strategic and Budgetary Assessments, May 18, 2010. As of October 1, 2012:
http://www.csbaonline.org/publications/2010/05/airsea-battle-concept

Woermann, Ashley, "Indonesian Military Expansion Strengthens China Partnership," *Future Directions International*, January 25, 2012. As of October 1, 2012:
http://www.futuredirections.org.au/publications/indian-ocean/29-indian-ocean-swa/
359-indonesian-military-expansion-strengthens-china-partnership.html

Xinhua, "China-Indonesia Joint Training for Special Forces Ends," *China.org.cn*, June 17, 2011. As of October 1, 2012:
http://www.china.org.cn/world/2011-06/17/content_22809317.htm

———, "China, Indonesia Eye for Closer Military Links," *China.org.cn*, January 16, 2012a. As of October 1, 2012:
http://www.china.org.cn/world/2012-01/16/content_24419726.htm

———, "Chinese Vice Premier Stresses Strategic Partnership with Indonesia," *People's Daily Online*, February 22, 2012b. As of October 1, 2012:
http://english.peopledaily.com.cn/90883/7736282.html